BIOLOGICAL TREATMENT OF SOLID WASTE

Enhancing Sustainability

BIOLOGICAL TREATMENT OF SOLID WASTE

Enhancing Sustainability

Edited by
Elena Cristina Rada, PhD

AAP | APPLE ACADEMIC PRESS

Apple Academic Press Inc. | Apple Academic Press Inc.
3333 Mistwell Crescent | 9 Spinnaker Way
Oakville, ON L6L 0A2 | Waretown, NJ 08758
Canada | USA

©2016 by Apple Academic Press, Inc.

First issued in paperback 2021

Exclusive worldwide distribution by CRC Press, a member of Taylor & Francis Group

No claim to original U.S. Government works

ISBN 13: 978-1-77463-587-2 (pbk)
ISBN 13: 978-1-77188-279-8 (hbk)

Library and Archives Canada Cataloguing in Publication

Biological treatment of solid waste : enhancing sustainability / edited by Elena Cristina Rada, PhD.

Includes bibliographical references and index.
ISBN 978-1-77188-279-8 (bound)
1. Sewage--Purification--Biological treatment. 2. Soil microbiology. 3. Compost. I. Rada, Elena Cristina, author, editor

TD755.B52 2015   628.3'5   C2015-903021-8

Library of Congress Cataloging-in-Publication Data

Biological treatment of solid waste : enhancing sustainability / Elena Cristina Rada, PhD, editor.

pages cm
Includes bibliographical references and index.
ISBN 978-1-77188-279-8 (alk. paper)
1. Sewage--Purification--Biological treatment. 2. Soil microbiology. 3. Compost. I. Rada, Elena Cristina, editor.

TD755.B4635 2015   628.3'5--dc23   2015016061

Apple Academic Press also publishes its books in a variety of electronic formats. Some content that appears in print may not be available in electronic format. For information about Apple Academic Press products, visit our website at **www.appleacademicpress.com** and the CRC Press website at **www.crc-press.com**

About the Editor

ELENA CRISTINA RADA, PhD

Elena Cristina Rada, PhD, earned her master's degree in Environmental Engineering from the Politehnica University of Bucharest, Romania; she received a PhD in Environmental Engineering and a second PhD in Power Engineering from the University of Trento, Italy, and the Politehnica University of Bucharest. Her post-doc work was in Sanitary Engineering from the University of Trento, Italy. She has been a professor in the Municipal Solid Waste master's program at Politehnica University of Bucharest, and has served on the organizing committees of "Energy Valorization of Sewage Sludge," an international conference held in Rovereto, Italy, and Venice 2010, an International Waste Working Group international conference. She also teaches seminars in the bachelor, master, and doctorate modules in the University of Trento and Padua and Politehnica University of Bucharest and has managed university funds at national and international level. Dr. Rada is a reviewer of international journals, a speaker at many international conferences, and the author or co-author of about a hundred research papers. Her research interests are bio-mechanical municipal solid waste treatments, biological techniques for biomass characterization, environmental and energy balances regarding municipal solid waste, indoor and outdoor pollution (prevention and remediation) and health, and innovative remediation techniques for contaminated sites and streams.

Contents

Acknowledgment and How to Cite

The editor and publisher thank each of the authors who contributed to this book. The chapters in this book were previously published elsewhere. To cite the work contained in this book and to view the individual permissions, please refer to the citation at the beginning of each chapter. Each chapter was carefully selected by the editor; the result is a book that looks at the sustainable treatment of solid waste from a variety of perspectives. The chapters included are broken into three sections, which describe the following topics:

- When considering the biologic treatment of solid wastes, microbial technologies are the foundation for all other considerations, while at the same time they also offer unique treatments of their own. The articles in chapters 1 through 4 were chosen to specifically examine the contributions and implications of various forms of microbial technology.
- Composting is one of the most ancient of all waste management systems, dating at least as far back as the first century of the Common Era. Today's composting treatments offer important solutions to solid waste problems. The articles in chapters 5 through 8 were selected to represent investigations into important aspects of this technology, including vermicomposting.
- Biodrying, the process by which biodegradable waste is rapidly heated through initial stages of composting to remove moisture from a waste stream, thus reducing its overall weight, is the next step in our considerations. Biodrying reduces the moisture content of waste without the need for supplementary fossil fuels, and with minimal electricity consumption—and it can take as few as 8 days to dry waste in this manner, which in turn reduces the cost of solid waste management. Biodrying is not yet a perfect technology, however, and it calls for ongoing research and development. The articles chosen for chapters 9 to 11 represent a sample of this most recent ongoing research.

List of Contributors

Norazlin Abdullah
Department of Process and Food Engineering, Faculty of Engineering, Universiti Putra Malaysia, 43400 UPM Serdang, Selangor, Malaysia

L. Alibardi
Department of Civil, Environmental and Architectural Engineering, University of Padova - Via Marzolo 9, 35131,Padova, Italy

Irini Angelidaki
Department of Environmental Engineering, Technical University of Denmark, Copenhagen, DK-2800, Kgs Lyngby, Denmark

Scott T. Bates
Department of Plant Pathology, University of Minnesota, Minneapolis, Minnesota, United States of America

Nyuk Ling Chin
Department of Process and Food Engineering, Faculty of Engineering, Universiti Putra Malaysia, 43400 UPM Serdang, Selangor, Malaysia

C. Collivignarelli
DICATAM, University of Brescia, via Branze 43, 25123 Brescia, Italy

R. Cossu
Department of Industrial Engineering, University of Padova - Via Marzolo 9, 35131, Padova, Italy

Xiaohu Dai
National Engineering Research Center for Urban Pollution Control, College of Environmental Science and Engineering, Tongji University, Shanghai, People's Republic of China

Davide De Francisci
Department of Environmental Engineering, Technical University of Denmark, Copenhagen, DK-2800, Kgs Lyngby, Denmark

Bin Dong
National Engineering Research Center for Urban Pollution Control, College of Environmental Science and Engineering, Tongji University, Shanghai, People's Republic of China

Noah Fierer
Cooperative Institute for Research in Environmental Sciences, University of Colorado, Boulder, Colorado, United States of America and Department of Ecology and Evolutionary Biology, University of Colorado, Boulder, Colorado, United States of America

F. Girotto
Department of Civil, Environmental and Architectural Engineering, University of Padova - Via Marzolo 9, 35131,Padova, Italy

Jingwei Jin
National Engineering Research Center for Urban Pollution Control, College of Environmental Science and Engineering, Tongji University, Shanghai, People's Republic of China

Panagiotis G. Kougias
Department of Environmental Engineering, Technical University of Denmark, Copenhagen, DK-2800, Kgs Lyngby, Denmark

Treu Laura
Department of Environmental Engineering, Technical University of Denmark, Copenhagen, DK-2800, Kgs Lyngby, Denmark

Jonathan W. Leff
Cooperative Institute for Research in Environmental Sciences, University of Colorado, Boulder, Colorado, United States of America

Gang Luo
Shanghai Key Laboratory of Atmospheric Particle Pollution and Prevention (LAP3), Department of Environmental Science and Engineering, Fudan University, Shanghai, 200433, China

Golden Makaka
Department of Physics, University of Fort Hare, Alice Campus, Alice 5700, Eastern Cape Province, South Africa

Sampson N. Mamphweli
Fort Hare Institute of Technology, University of Fort Hare, Alice Campus, Alice 5700, Eastern Cape Province, South Africa

Christy E. Manyi-Loh
Fort Hare Institute of Technology, University of Fort Hare, Alice Campus, Alice 5700, Eastern Cape Province, South Africa and Applied and Environmental Microbiology Research Group (AEMREG), Department of Biochemistry and Microbiology, University of Fort Hare, Alice Campus, Alice 5700, Eastern Cape Province, South Africa

Edson L. Meyer
Fort Hare Institute of Technology, University of Fort Hare, Alice Campus, Alice 5700, Eastern Cape Province, South Africa

E. Minciuc
University Politehnica of Bucharest, Romania

Mohd Noriznan Mokhtar
Department of Process and Food Engineering, Faculty of Engineering, Universiti Putra Malaysia, 43400 UPM Serdang, Selangor, Malaysia

Deborah A. Neher
Department of Plant and Soil Science, University of Vermont, Burlington, Vermont, United States of America

M. Norisor
Power Engineering Faculty, University Politehnica of Bucharest

Anthony I. Okoh
Applied and Environmental Microbiology Research Group (AEMREG), Department of Biochemistry and Microbiology, University of Fort Hare, Alice Campus, Alice 5700, Eastern Cape Province, South Africa

Jayakumar Pathma
Department of Biotechnology School of Life Sciences, Pondicherry University, Kalapet, Puducherry, 605014, India

R. Patrascu
Faculty of Power Engineering, University Politehnica of Bucharest, 060032 Bucharest, Romania

A. Perteghella
DICATAM, University of Brescia, via Branze 43, 25123 Brescia, Italy

Elena Cristina Rada
Civil and Environmental Department, University of Trento, via Mesiano 77, 38100, Trento, Italy

Marco Ragazzi
Civil and Environmental Department, University of Trento, via Mesiano 77, 38100, Trento, Italy

Natarajan Sakthivel
Department of Biotechnology School of Life Sciences, Pondicherry University, Kalapet, Puducherry, 605014, India

Michael Simon
Fort Hare Institute of Technology, University of Fort Hare, Alice Campus, Alice 5700, Eastern Cape Province, South Africa

Farah Saleena Taip
Department of Process and Food Engineering, Faculty of Engineering, Universiti Putra Malaysia, 43400 UPM Serdang, Selangor, Malaysia

D. Tutica
Academy of Romanian Scientists, Romania

M. Vacchari
DICATAM, University of Brescia, via Branze 43, 25123 Brescia, Italy

Thomas R. Weicht
Department of Plant and Soil Science, University of Vermont, Burlington, Vermont, United States of America

Jing Yi
National Engineering Research Center for Urban Pollution Control, College of Environmental Science and Engineering, Tongji University, Shanghai, People's Republic of China

Xinyu Zhu
Department of Environmental Engineering, Technical University of Denmark, Copenhagen, DK-2800, Kgs Lyngby, Denmark

Introduction

Throughout most of history, humans had very little problem deciding what to do with left-over substances and objects for which they had no use. Waste material produced during pre-modern times was mainly ashes, vegetable refuse, and human and animal waste. These could all be released back into earth with minimum environmental impact. Objects made from metal—tools and jewelry, for the most part—were expected to last a lifetime, and then were passed on through generations.

As dense population centers grew, however, waste disposal became a problem, one that was exacerbated by industrialization. Solid wastes were often dumped into waterways, and diseases such as cholera spread through the polluted water. By the mid-nineteenth century, the problem had become so severe that it became a public debate that led to the first sanitation legislation.

Incinerators were the most common means of waste disposal through the rest of the nineteenth century and into the twentieth century. In the twentieth century, landfills became another option. Both forms of waste management had serious potential for polluting the land, air, and water around urban waste management plants.

Today, the development of sustainable solutions for waste management is one of society's great challenges. In this book, we will discuss biological treatment methods, including microbial technologies, composting, and biodrying.

Food waste is a part of the solid waste challenge. In chapter 1, Girotto and colleagues discuss several possibilities that are currently available for food waste. They offer a management hierarchy that begins with prevention and reduction of food waste. The next step is salvaging food for the poor from food waste, followed by reusing food waste to feed animals. Next in line are industrial uses (biofuels and plastic production, for example). Once these most sustainable options have been exhausted, composting is next in line. Incineration and landfill should be the last option considered—and hopefully, having passed through all the other layers of the hierarchy, the solid waste that reaches the end will be greatly reduced. The benefits of this are obvious.

Microbes play a significant role in two levels of this hierarchy: industrial uses and composting. Recently, numerous studies have analyzed microbial communities for their potential to play a part in sustainable management of solid waste. Pyrosequencing has gained increasing attention as next-generation technology for studying microbial diversity. In chapter 2, Jing Yi and colleagues apply this to various environmental samples, such as source waste, membrane filtration systems, and soil, in order to compare the microbial community structures in anaerobic digestion of food waste at different organic loading rates.

Overall, anaerobic digestion reduces biomass wastes and mitigates a wide spectrum of solid wastes' impact on the environment. It improves sanitation, helps to limit air and water pollution, and reduces greenhouse gas emissions. It also provides a high-quality nutrient-rich fertilizer and can yield energy in the form of biogas. Particularly in developing countries, biogas offers important solutions to societal demands. It can be used as fuel for cooking, lighting, and heating; it reduces the demand for wood and charcoal for cooking, which helps preserve forests and other natural environments, and it produces less indoor pollution when used than wood and charcoal do. In Western countries, biogas is converted to electricity and heat.

In chapter 3, Manyi-Loh and her colleagues offer a comprehensive description of anaerobic digestion as a means to resolve or mitigate the current dangers from animal waste disposal. They emphasize certain types of biodigesters (microbial communities), and conclude that anaerobic digestion of animal manure is a strong option for either safely reusing wastes or else transforming them into valuable materials and energy. The decomposition process reduces wastes' oxygen demand, destroys pathogenic microbes that pose health risks, destroys volatile fatty acids, and reduces greenhouse gas emissions. Ultimately, as already mentioned, it generates biogas and high-quality nutrient-rich fertilizer.

Based on these considerations, the objective of Gang Luo and colleagues in chapter 4 was to understand the role of stochastic factors and disturbance in the steady-state microbial community and its function in biogas reactors. They used three replicate biogas reactors to treat cattle manure in order to determine whether similar microbial communities would be achieved at steady states if the reactors were operated under the same conditions. Temperature stability is an essential factor for the biogas process, but biogas reactors are subject to temperature fluctuations due to mechanical errors and breakdowns. Gang Luo and colleagues therefore intentionally fluctuated the temperature in the three reactors in order to determine how much the temperature disturbance would alter the steady-state microbial community. They also monitored the reactors' performances as to biogas production, pH, and total volatile fatty acids. They found that similar steady-state process performance and microbial community profiles were achieved in the three reactors, suggesting that stochastic factors have a minor role in shaping the profile of the microbial community composition and activity in biogas reactor. Temperature disturbance, however, played a very important role in the microbial community composition as well as process performance. After temperature disturbances, all three reactors had increased methane yields and decreased volatile fatty acids concentrations. The authors also observed new steady-state microbial community profiles in all the biogas reactors after a temperature disturbance.

In chapter 5, we turn to composting, a form of solid waste management that also offers the opportunity to produce fertilizer for farming. Authors Collivignarelli and colleagues compare two compost heaps and discuss temperature trends and physical-chemical analyses. Their results were intended for practical application to a full-scale plant that treats 6 tons of organic waste per month. Their test on the pilot plant revealed that heap overturning frequencies can influence the composting process in terms of duration, temperature, and final product quality. More frequent heap overturning (every 5 days) avoided possible temperature peaks. Just as temperature fluctuations can have a negative impact on bioreactors, as discussed in chapter 4, they can also impede the microbial communities in compost. Overturning can also reduce the biological process duration and therefore also reduce the material's treatment time in the compost plant, allowing the plant to handle a larger quantity of daily waste. The authors found that too much overturning, however, can cause an excessive volatilization of organic parameters. Even then, the quality of compost produced was good, since there was no heavy metal pollution. Collivignarelli and

colleagues conclude that the pilot tests' results represented a good starting point for the full-scale composting plant, while at the same time they
called for ongoing monitoring and adjustments in order to achieve good
quality compost production

Microbial communities in compost are abundant and diverse, influenced by both recipe and post-thermophilic treatment. Chapter 6, by Neher and her colleagues, is a comprehensive assessment of the bacteria and
fungi associated with compost and the influence of both composting recipe
and process on the structure of microbial communities. Of particular interest is their analysis of the ways in which communities change through time
when compost is produced on a commercial-scale. They note that composition starts similarly after the thermophilic phase, but then shifts dynamically through time. Economic considerations encourage commercial
composters to speed up the composting process, which has contributed to
a focus on the effectiveness of the thermophilic phase. The authors note,
however, that the curing phase offers a substrate and climate conducive for
microbial recolonization, which can be accomplished either by inoculating
post-thermophilic compost or preparing a palatable substrate that provides
a competitive advantage for microbial colonization, slow-release fertility,
and plant growth. Neher and her colleagues call for future research building on their results, in order to determine which recipe and post-thermophilic phase will best promote the agricultural goals of weed management,
disease suppression, and plant growth promotion.

Composting has various practical requirements. One of these is a bulking agent. In chapter 7, Abdullah and colleagues study the effects of two
common bulking media: newspaper and onion peels. They found that onion peels were more suitable; at a smaller waste load, compost maturity
was attained more quickly with the onion peels than with the newspaper.

Vermicomposting is a cost-effective and eco-friendly waste management technology that has many advantages over traditional thermophilic
composting. Vermicomposts are excellent sources of biofertilizers that
improve the physiochemical and biological properties of agricultural soil.
Vermicomposting amplifies the diversity and population of beneficial microbial communities. In chapter 8, Pathma and Sakthivel conclude that
vermicomposting is beneficial in numerous ways, including soil reclamation, soil fertility, plant growth, and control of pathogens, pests, and

nematodes. They point to this method as one of the most promising for sustainable agriculture.

Last, we consider biodrying. In chapter 9, we outline the advantages of this technology. These include reducing the cost per unit of thermal energy, as well as reducing the global impact of fossil fuels. We recommend that those nations complying with the Kyoto protocol should consider biodrying technology as a means to achieve their targets.

In Chapter 10, Minciuc and his colleagues focus on biological degradation of organic matter of animal origin aimed to energy exploitation. They discuss the formation of combustible gases due to the biological degradation. This is a process that takes place naturally on our planet. It is how natural gas was formed, and now we have the option of replicating that process. The authors present two solutions in order to use the animal waste for biogas production and its utilization for energy generation: electricity is used either in the plant or is sold to the company managing the national grid. They present also an economic balance and conclude that within the context of sustainable development, where economic, energy, and environmental issues must all be addressed simultaneously, the proposed technology depends on the input material that influences the biogas production.

Finally, in chapter 11, we examine the ways in which biodrying exhausted grape marc can be an improved alternative to thermal drying. Thermal drying generally requires an integrated and centralized plant of thermal pretreatment and combustion. Grape marc biodrying could open an option for decentralized pretreatment of this organic substrate before a centralized combustion.

Our world can no longer afford to consider waste as something that can be discarded with no regard for future use. Instead, if addressed correctly through policy and practice, solid waste can become a valuable resource. With rational and consistent waste management practices come opportunities to reap a range of benefits. Those benefits include economic, social, environmental factors. Biological treatment of solid wastes can help build a better world for future generations, one that has a healthier economy, a healthier society, and a healthier environment

Elena Cristina Rada

Food waste can be defined as material intended for human consumption that is instead discharged, lost, degraded or contaminated. In Chapter 1, Girotto and colleagues propose several solutions for the proper management of this food waste. Such solutions can be prioritised similarly to the waste management hierarchy. First steps are minimisation and use to feed poor. Food waste can then be used in industrial processes for the production of biofuels or biopolymers. Further steps foresee the recovery of nutrients and the fixation of carbon by composting. Final and less desirable solutions are incineration and landfilling.

The total solids content of feedstocks affects the performances of anaerobic digestion and the change of total solids content will lead the change of microbial morphology in systems. In order to increase the efficiency of anaerobic digestion, it is necessary to understand the role of the total solids content on the behavior of the microbial communities involved in anaerobic digestion of organic matter from wet to dry technology. In Chapter 2, Yi and colleagues compared the performances of mesophilic anaerobic digestion of food waste with different total solids contents from 5% to 20% and investigated the microbial communities in reactors were using 454 pyrosequencing technology. Three stable anaerobic digestion processes were achieved for food waste biodegradation and methane generation. Better performances mainly including volatile solids reduction and methane yield were obtained in the reactors with higher total solids content. Pyrosequencing results revealed significant shifts in bacterial community with increasing total solids contents. The proportion of phylum *Chloroflexi* decreased obviously with increasing total solids contents while other functional bacteria showed increasing trend. *Methanosarcina* absolutely dominated in archaeal communities in three reactors and the relative abundance of this group showed increasing trend with increasing total solids contents. These results revealed the effects of the total solids content on the performance parameters and the behavior of the microbial communities involved in the anaerobic digestion of food waste from wet to dry technologies.

With an ever increasing population rate; a vast array of biomass wastes rich in organic and inorganic nutrients as well as pathogenic microorganisms will result from the diversified human, industrial and agricultural activities. In Chapter 3, Manyi-Loh and colleagues posit anaerobic digestion as one of the best ways to properly handle and manage these wastes. Animal wastes have been recognized as suitable substrates for anaerobic digestion process, a natural biological process in which complex organic materials are broken down into simpler molecules in the absence of oxygen by the concerted activities of four sets of metabolically linked microorganisms. This process occurs in an airtight chamber (biodigester) via four stages represented by hydrolytic, acidogenic, acetogenic and methanogenic microorganisms. The microbial population and structure can be identified by the combined use of culture-based, microscopic and molecular techniques. Overall, the process is affected by bio-digester design, operational factors and manure characteristics. The purpose of anaerobic digestion is the production of a renewable energy source (biogas) and an odor free nutrient-rich fertilizer. Conversely, if animal wastes are accidentally found in the environment, it can cause a drastic chain of environmental and public health complications.

The microbial community in a biogas reactor greatly influences the process performance. However, only the effects of deterministic factors (such as temperature and hydraulic retention time (HRT)) on the microbial community and performance have been investigated in biogas reactors. Little is known about the manner in which stochastic factors (for example, stochastic birth, death, colonization, and extinction) and disturbance affect the stable-state microbial community and reactor performances. In Chapter 4, Luo and colleagues ran three replicate biogas reactors treating cattle manure to examine the role of stochastic factors and disturbance in shaping microbial communities. In the triplicate biogas reactors with the same inoculum and operational conditions, similar process performances and microbial community profiles were observed under steady-state conditions. This indicated that stochastic factors had a minor role in shaping the profile of the microbial community composition and activity in biogas reactors. On the contrary, temperature disturbance was found to play an important role in the microbial community composition as well as process performance for biogas reactors. Although three different temperature disturbances were applied to each biogas reactor, the increased methane yields (around 10% higher) and decreased volatile fatty acids (VFAs) concentrations at steady state were found in all three reactors after the temperature disturbances. After the temperature disturbance, the biogas reactors were brought back to the original operational conditions; however, new steady-state microbial community profiles were observed in all the biogas reactors. The present study demonstrated that temperature disturbance, but not stochastic factors, played an important role in shaping the profile of the microbial community composition and activity in biogas reactors. New steady-state microbial community profiles and reactor performances were observed in all the biogas reactors after the temperature disturbance.

Organic solid waste and the current lack of sound treatment for its valorisation are aspects of high priority for municipal decision makers in developing countries. Despite the remarkable use of composting processes in the field in developing countries, the availability of databases and information like those of developed countries is still lacking in literature. Thus, this lack represents constrains towards the sustainable implementation of the composting. The research and monitoring activities conducted on Maxixe composting pilot plant were carried out in order to obtain valuable and useful data and information for their further adoption on the full scale composting plant. Chapter 5, by Collivignarelli and colleagues, shows how different overturning frequencies influence the composting process of two heaps with the same initial characteristics, comparing also chemical analyses of the final compost. Moreover this paper is aimed to spread out results from field work and contributing at the same time to fill the current gap of real field data concerning composting process in developing countries.

Compost production is a critical component of organic waste handling, and compost applications to soil are increasingly important to crop production. However, we know surprisingly little about the microbial communities involved in the composting process and the factors shaping compost microbial dynamics. In Chapter 6, Neher and colleagues used high-throughput sequencing approaches to assess the diversity and composition of both bacterial and fungal communities in compost produced at a commercial-scale. Bacterial and fungal communities responded to both compost recipe and composting method. Specifically, bacterial communities in manure and hay recipes contained greater relative abundances of Firmicutes than hardwood recipes with hay recipes containing relatively more Actinobacteria and Gemmatimonadetes. In contrast, hardwood recipes contained a large relative abundance of Acidobacteria and Chloroflexi.

Fungal communities of compost from a mixture of dairy manure and silage-based bedding were distinguished by a greater relative abundance of Pezizomycetes and Microascales. Hay recipes uniquely contained abundant Epicoccum, Thermomyces, Eurotium, Arthrobotrys, and Myriococcum. Hardwood recipes contained relatively abundant Sordariomycetes. Holding recipe constant, there were significantly different bacterial and fungal communities when the composting process was managed by windrow, aerated static pile, or vermicompost. Temporal dynamics of the composting process followed known patterns of degradative succession in herbivore manure. The initial community was dominated by Phycomycetes, followed by Ascomycota and finally Basidiomycota. Zygomycota were associated more with manure-silage and hay than hardwood composts. Most commercial composters focus on the thermophilic phase as an economic means to insure sanitation of compost from pathogens. However, the community succeeding the thermophilic phase begs further investigation to determine how the microbial dynamics observed here can be best managed to generate compost with the desired properties.

To prevent the interruption of the carbon cycle by the disposal of waste to landfills, organic kitchen waste requires proper treatment such as composting to reduce its uncontrolled degradation on disposal sites and subsequent greenhouse gases, odour emissions and nutrient losses. Chapter 7, by Abdullah and colleagues, investigated the effects of bulking agent, newspaper and onion peels, composting waste load sizes of 2 and 6 kg, or the use of starter culture on kitchen-waste composting consisting of nitrogen-riched substrates, vegetable scraps and fish processing waste in an in-vessel system. The optimised formulation of kitchen waste mixture was used for a 30-day composting study, where the temperature profiles were recorded and the carbon-to-nitrogen ratios were measured as an indication of compost maturity. The kitchen-waste composting process was conducted in parallel in two fabricated kitchen waste composters. It was found that the onion peels were more suitable in producing matured compost where the carbon-to-nitrogen ratio reduced to 10 within 16 days of composting. A smaller kitchen waste load size of 2 kg gave a shorter composting time by half when compared to the 6 kg. The use of a microbial cocktail consisting seven types of bacteria and eight types of fungi isolated from soils as a starter culture for this kitchen-waste composting did not show advantages in accelerating the composting process. The results suggest that the in-vessel kitchen-waste composting can be efficient with a minimal load of about 2 kg using onion peels without additional starter culture.

Vermicomposting is a non-thermophilic, boioxidative process that involves earthworms and associated microbes. This biological organic waste decomposition process yields the biofertilizer namely the ver-micompost. Vermicompost is a finely divided, peat like material with high porosity, good aeration, drainage, water holding capacity, microbial activity, excellent nutrient status and buffering capacity thereby result-ing the required physiochemical characters congenial for soil fertility and plant growth. Vermicompost enhances soil biodiversity by promot-ing the beneficial microbes which inturn enhances plant growth directly by production of plant growth-regulating hormones and enzymes and indirectly by controlling plant pathogens, nematodes and other pests, thereby enhancing plant health and minimizing the yield loss. Due to its innate biological, biochemical and physiochemical properties, Pathma and Sakthivel argue in Chapter 8 that vermicompost may be used to promote sustainable agriculture and also for the safe management of ag-ricultural, industrial, domestic and hospital wastes which may otherwise pose serious threat to life and environment.

In the frame of municipal solid waste management, one of the avail-able options is based on the generation of refuse derived fuel (RDF) for industrial use. To this end, in Chapter 9 Rada and Raguzzi propose bio-drying for a decentralized management of waste: a few satellite plants could generate RDF for a centralized use. This decentralization could be based on a small scale plant for local RDF exploitation. To this concern in the sector there is not yet an approach useful for assessing the viability of bio-drying when a factory requiring heat wants to generate exactly the amount of RDF to be used in its own dedicated RDF burner. In the pres-ent paper some criteria for assessing the viability of this strategy at a very small scale are presented. Co-generation is not taken into account because of the unfavourable scale. The method takes into account many aspects: characteristics of the factory, of the waste, of the RDF that can be gener-ated and also of the local economy.

In Chapter 11, Rada and Raguzzi carried out an experimentation to study the behavior of exhausted grape marc during the bio-drying process. This process was chosen as an alternative to the typical grape marc thermal drying approach. The aim was to reduce the moisture level thanks to the biological exothermal reactions, and to increase the energy content in the bio-dried grape marc. The target was the generation of a product interesting for energy options. For the development of the research, a biological pilot reactor and a respirometric apparatus were used. Results demonstrated that bio-drying can decrease the water content saving the original energy content. The final material could be assumed like a Solid Recovered Fuel, class 5:1:1 with a very low potential rate of microbial self heating.

PART I

MICROBIAL TECHNOLOGIES

CHAPTER 1

Management Options of Food Waste: A Review

F. GIROTTO, L. ALIBARDI, AND R. COSSU

1.1 INTRODUCTION

The management of food waste is currently a growing problem in the waste management field and the search for sustainable solutions represents a challenge not only for this specific sector but also for the agricultural and industrial sectors and requires the involvement of retailers and consumers.

The Food and Agriculture Organization of the United Nations (FAO) provides a definition for food waste. It is the food loss occurring during the retail and final consumption stages due to the behaviour of retailers and consumers, where the term food loss itself measures the decrease in edible food mass (excluding inedible parts and seeds) throughout the part of the supply chain that specifically leads to edible food for human consumption, that is, loss at the production, postharvest and processing stages (FAO, 1983).

Girotto F, Alibardi L, and Cossu R. "Management Options of Food Waste: A Review" SUM 2014, Second Symposium on Urban Mining Bergamo, Italy; 19 – 21 May 2014. *CISA Publisher (2014).* *Used with permission from the publisher.*

A deep analysis of a whole Food Supply Chain (FSC) system can highlight the fact that the production of waste material (organic waste or food waste) regards all sectors involved in the production, distribution and consumption of food.

In the definition provided by FAO and also in those from other sources (Smill, 2004; Stuart, 2009; Gustavsson et al., 2011), the concept of food waste is wide and regards not only the production of food waste at household level but also the decrease in edible food mass throughout the human FSC. In this food waste definition therefore, all those materials that are edible food but that are wasted for several reasons, are considered.

A FSC starts with food production from the agricultural sector where both framing and husbandry produce waste or subproducts that can be both organic waste (i.e. cornstolk, manure) or food waste (i.e. low quality fruits or vegetable, damaged productions left in the field, good products or co-products but with low or no commercial value). The edible (or potentially edible) materials are considered food waste. The further food processing and manufacturing industrial sector can produce food losses and food waste during the entire production phase for several reasons. Damage during transport or non-appropriate transport systems, problems during storage, losses during processing or contaminations, inappropriate packaging, are some examples of food waste production in the agro-industrial sector. The retailing system and markets also produce food waste, due to problems in conservation or handling, lack in cooling/cold storage. The production of food waste from the final consumer is due to over- or non appropriate purchasing, bad storage conditions, over-preparation, portioning and cooking as well as confusion between the terms "best before" or "use by" dates. (Papargyropoulou et al., 2014). The generation of food waste at household level is influenced by several interconnected factors, the main ones being socio-demographic characters of household, consumption behaviour and food patterns (Glanz and Schneider, 2009).

Food waste production has impacts at environmental, social and economical levels. From an environmental point of view, food waste contributes to Green House Gas (GHG) emissions during final disposal in landfills (uncontrolled methane release) and during activities associated to food production, processing, manufacturing, transportation, storage and distribution. Other environmental impacts associated to food waste are natural resource depletion in terms of soil, nutrients, water and energy, disruption of biogenic cycles due to intensive agricultural activities and all other characteristics impacts in any step of the FSC. Social impacts of food waste can be ascribed at ethical and moral dimension within the general concept of global food security. Economical impacts are due to the costs related to food westage and their effects on farmers and consumers incomes (Lipinski et al., 2013; Papargyropoulou et al., 2014).

This paper reviews the management options reported and discussed in the scientific literature with the final aim of providing a clear and complete vision on the magnitude of food waste production and the possibilities currently available for food waste management and their state of the art.

1.2 PRODUCTION OF FOOD WASTE

The Food and Agriculture Organization of the United Nations (FAO, 2004) estimates that 32% of all food produced in the world was lost or wasted in 2009. While 870 million people are reported to be chronically undernourished, about 1.3 billion tons/year, that is one third of the food produced for human consumption, is wasted globally (Kojima and Ishikawa, 2013). In Europe, food waste production is estimated at about 90 million tons annually (EC, 2013), in the United States it is estimated around 27% of the industrially produced food (Garcia, 2011) and in Japan it is about 21 million tons in 2010 (Kojima and Ishikawa, 2013). Quested et al., (2013) reported a production of food waste at household level of 160 kg per year in United Kingdom (UK), representing 12% of the food and drink entering a home and 30% of the general waste stream from UK household. The order of magnitude of food waste production is consistent and is not limited to developed countries. Gustavson at al. (2011) reported data on the food waste production around the World and results are summarised in Table 1.

Food waste production in industrialised countries has similar order of magnitudes as in developing countries (DCs) (EC, 2014). Nevertheless industrialized and developing countries differ substantially as reported in Table 1. In the latter, more than 40% of food losses occur at the postharvest and processing stages, while in the former, about 40% of losses occur at the retail and consumer levels and, on a per-capita basis, much more food is wasted in the industrialized world than in developing countries (Gustavsson et al., 2011).

TABLE 1: Average annual food waste production per person in different areas of the World (Modified from Gustavson et al., 2011)

Area	Food waste production (kg/person/year)		
	Total	Production and retail stage	Consumer's stage
Europe	280	190	90
North America and Oceania	295	185	110
Industrialized Asia	240	160	80
sub-Saharan Africa	160	155	5
North Africa, West and Central Asia	215	180	35
South and Southeast Asia	125	110	15
Latin America	225	200	25

The causes of food losses and waste in low-income countries are mainly connected to financial, managerial and technical limitations in harvesting techniques, storage and cooling facilities in difficult climatic conditions, infrastructure, packaging and marketing systems. Given that many smallholder farmers in DCs live on the margins of food insecurity, a reduction in food losses could have an immediate and significant impact on their livelihoods. The food supply chains in DCs need to be strengthened, encouraging small farmers to organize, diversify and upscale their production and marketing. Investments in infrastructure, transportation, food industries and packaging industries are also required. Both the public and private sectors have a role to play in achieving this.

The causes of food losses and waste in medium/high-income countries mainly relate to consumer behaviour as well as to a lack of coordination between different actors in the supply chain. Farmer-buyer sales agreements may contribute to quantities of farm crops being wasted. Food can be wasted due to quality standards, which reject food items not perfect in shape or appearance. At the consumer level, insufficient purchase planning and expiring "best before dates" also cause large amounts of waste, in combination with the careless attitude of those consumers who can afford to waste food. Food waste in industrialized countries can be reduced by raising awareness among food industries, retailers and consumers. This inevitably also means that huge amounts of the resources used in food production are used in vain, and that the greenhouse gas emissions caused by production of food that gets lost or wasted are also emissions in vain.

According to Venkat (2011), at least 123 million metric tonnes of CO_2 emissions are added to the atmosphere each year from the production, transport and disposal of the uneaten food. This translates to over 13% of all food-related emissions in the US and about 1.5% of total US emissions, and most of these emissions come from the production stage. The cost estimation indicated that consumers and businesses waste nearly $200 billion worth of raw food commodities annually (Venkat, 2011). Including in the evaluation other food commodities that are produced and wasted in smaller quantities, together with other emission sources such as additional processing, packaging and cooking, the overall cost and climate change impact of food waste would be higher than the reported preliminary calculation. This should stand for an additional incentive to push the food industry to reduce food waste production in order to gain benefits on both the financial and environmental fronts. A much larger share of the food waste burden rests with consumers, counting household food waste and all of the plate loss away from home. Since consumers do not have the tools and systems to manage their inventories and food preparation, putting a dent in this part of the waste will require a combination of education, the availability of more optimal portion sizes away from home, innovative food packaging/preservation techniques. Cutting household waste food would be helpful to go one step further towards reducing food-related emissions besides representing an efficient way for families to save money.

1.3 FOOD WASTE MANAGEMENT HIERARCHY

Similarly to the Waste Management Hierarchy introduced in Europe in 1989 (European Parliament Council, 1989), based on a hierarchy of solutions of five distinct steps (waste prevention, reuse, recovery and recycling of materials, energy recovery and safe landfilling of residues) and often graphically represented by a reverse triangle (Cossu, 2009), the Environmental Protection Agency (EPA, 2014) defined the following hierarchy concept in relation to food waste management: source reduction, feed hungry people, feed animals, industrial uses, composting, incineration or landfilling. A graphical representation is reported in Figure 1, as a reverse triangle, where the larger is the bar of the step, the more viable is the management option.

The first step in the reduction of food waste production starts from the issue of the undesirable food surplus, the prevention of over-production and over-supply of food beyond human nutritional needs (Papargyropoulou et al., 2014).

The first step of the hierarchy also involves food waste reduction throughout the entire FSC. This focus has to be different from country to country as highlighted by the work of Gustavsson et al. (2011). In developed Countries, food waste prevention should focus on the consumer's behaviours at household level while in developing Countries it should focus more on the retail and distribution system. Very close to food waste prevention are the issues of food security and the utilisation of food surplus to satisfy the nutritional needs of poor people.

The next steps of the hierarchy are the utilisation of food waste to feed animals and in industrial sectors. Within industrial utilisation, several options can be described varying from the use of food waste for energy production by anaerobic digestion (bio-hydrogen or bio-methane productions) to the production of specific chemical compounds as precursors for plastic materials production, chemical or pharmaceutical applications.

The composting step has the main aim of nutrients recovery and carbon sequestration by humic substances formation. The last, less desirable step is represented by landfilling or incineration. It is well known that biodegradable organic material represents the main source of adverse environmental impacts and risks in traditional landfilling (odours, fires, VOC's, groundwater contamination by leachate, global climate changes, etc.) (i.a. Manfredi et al., 2010; Thomsen et al., 2012; Beylot et al., 2013) while thermal treatment, even though providing energy recovery, is limited by the low heating values of organic waste (Nelles et al., 2010). For these reasons, such options represent the less desired ones.

Source reduction

Feed poor

Feed animals

Industrial uses

Composting

Incineration
or
landfilling

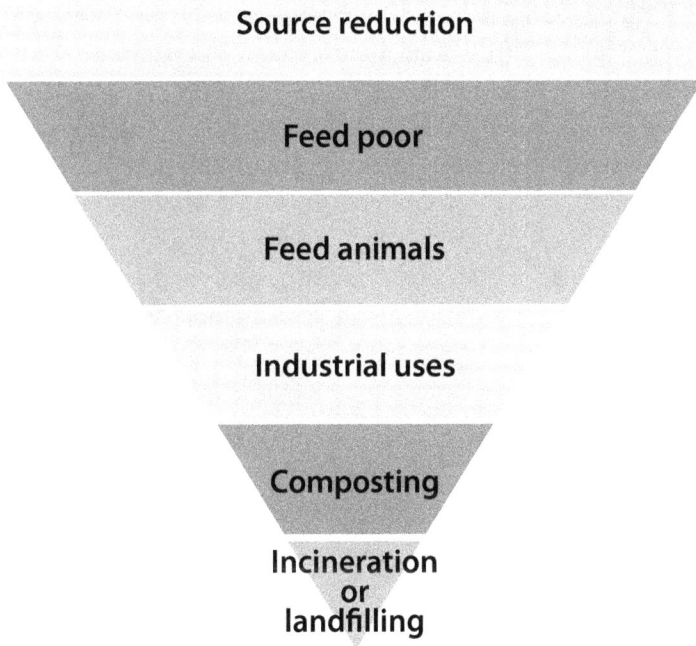

FIGURE 1: Food waste management hierarchy pyramid (EPA, 2014).

1.3.1 PREVENTION

One way of dealing with food waste is the reduction of its production. This attitude has been promoted by campaigns from advisory and environmental groups and by concentrated media attention on the subject. Several papers have analysed the behaviour of companies and citizens in developed countries at different levels (household, restaurant, retail) to assess the governing factors influencing waste of food products (Glanz and Schneider, 2009; Schneider and Lebersorger, 2009; Silvennoinen et al., 2012; Quested et al., 2013; Katajajuuri et al., 2014; Garrone et al., 2014; Graham-Rowe et al., 2014; Mena et al., 2014).

To prevent food waste production, different types of actions carried out at several levels are required. The retail system can cause food waste production in various phases of the food product distribution and purchase: damages during transport or non appropriate transport systems, problems during intermediate storages, losses during processing or contaminations, inappropriate packaging, problems in conservation or handling, lack in cooling/cold storage.

A huge impact on food waste production is held by the behaviour of consumers at household level. Over- or non appropriate purchasing, bad storage conditions, over-preparation, portioning and cooking as well as confusion between the terms "best before" or "use by" dates are some of the main factors affecting food loss. This behaviour is influenced by several interconnected factors, the main ones being socio-demographic characters of household, consumption behaviour and food patterns. The barriers to surpass to reach food loss minimisation at household level can also involve emotional and psychological aspects. One example is the wish to be "good provider" in terms of healthy and abundant food for the family. Lack in food may produce a sense of inappropriate ability to take care of the needs of the family thus pushing the purchase of more items even though not necessary ("bread every day in the table" or "never an empty fridge"). Another example is the avoidance of frequent trips to shops, thus pushing people to buy more food products to prevent or avoid inconveniences. A general lack in awareness of the amount of food waste generated at household level can have a strong impact on food waste production, due to the fact that small quantities thrown away a bit at a time with other waste does not provide the proper order of magnitude of the problem to consumers (Graham-Rowe et al., 2014).

1.3.2 FEEDING THE POOR

Organizations can donate safe and healthy food to food banks or food rescue organizations. This simple effort would lead to a double achievement: the reduction of food sent to landfills and the feeding of those in need. In the US, for example, the *"Emerson Good Samaritan Food Donation Act"* was created to encourage food donation to non-profits by minimizing liability (Culver, 2013). Donating wholesome and edible food can also claim tax benefits (EPA, 2014).

More and more widespread is garbage picking: the practice of sifting through commercial or residential waste to find items that have been discarded by their owners, but may prove useful to the garbage picker. Garbage picking may take place in dumpsters or in landfills. Nowadays, unfortunately, because of the globally critical economical situation, many people find themselves compelled to be "dumpster divers" also for food as well as clothing, furniture, etc. Dumpster diving is practiced differently in developed countries and developing countries. In many developing countries, food is rarely thrown away unless it is rotten. As already stated, in fact, the highest percentage of food loss occurs at the postharvest and processing stages. In developed nations, like the United States, where 40–50% of food is wasted, the trash contains a lot more food to gather (Harrison, 2004). In regions where people practice dumpster diving, food waste is reduced. However, it can pose a health risk to these people and there may also be questions of legality.

1.3.3 FEEDING ANIMALS

The feeding of food scraps to animals is, historically, the most common way of dealing with household food waste. From the end of the 19th century through the middle of the 20th Century, many municipalities collected food waste (called "garbage" as opposed to "trash") separately. This was typically disinfected by steaming and fed to pigs, either on private farms or in municipal piggeries. Now feeding scraps to worms that produce soil as a by-product is also a widespread practise called vermicomposting.

1.3.4 INDUSTRIAL USES

1.3.4.1 BIOFUEL PRODUCTION

Currently more and more effort is invested in the fourth step of the food waste management hierarchy. The definition of effective and stable ways to obtain biofuel and bio-plastic from waste food represents in fact one of the most interesting goals of this century. This possibility could demonstrate benefits from an environmental point of view due to the reduction of methane gas emissions from landfills and the preservation of natural resources such as coal and fossil fuels, from a social point of view because there would be no food vs. fuel competition to obtain bioethanol or biodiesel, and from an economical point of view thanks to costs saving connected to surplus food production and specific investments to grow no food crops dedicated to biofuel or bioplastic production.

The industrial concept behind food waste utilisation is the biorefinery. Similarly to the way that oil refineries convert petroleum into fuels and ingredients for hundreds of consumer products, biorefineries convert corn, sugar cane, and other plant-based material into a range of ingredients for bio-based fuels and other products.

Improving energy security and mitigating climate changes are among the most important bioenergy drivers in most countries. Concerns about the negative environmental effects due to the utilization of fossil fuels (GHG emissions into atmosphere and global warming effects), rising prices of crude oil and increasing demand for transportation fuel, are the major constraints in the economic development of many nations that could stimulate investment, research and industrial application of the biorefinery concept. Of course, much attention has to be paid in order to develop the right techniques and strategies to obtain the desired substances without implying the increase of the prices of edible products and without causing land use competition between food and biofuels. Biofuels and bio-based materials will only be beneficial if they are produced in a sustainable way with both biodiversity and "food vs. fuel"-debate in mind (Nigam and Singh, 2011; Refaat, 2012).

Biodiesel can be defined as fatty acid alkyl esters (methyl/ethyl esters) of short-chain alcohols and long-chain fatty acids derived from natural biological lipid sources like vegetable oils or animal fats, which have had their viscosity reduced using a process called transesterification and can be used in conventional diesel engines and distributed through existing fuel infrastructure. Any fatty acid source may be used to prepare biodiesel (Refaat, 2012). Thus, any animal or plant lipid should be a ready substrate for the production of biodiesel. However the use of edible vegetable oils and animal fats for biodiesel production has recently been of great concern because they compete with food materials while the use of nonedible vegetable oils for biodiesel production is also questionable because the growing of crops for fuel wastes land, water, and energy resources vital for the production of food for human consumption. The conclusion that can be derived is that the use of waste oil for the production of biodiesel is the most realistic and effective. The new process technologies developed during recent years have made it possible to produce biodiesel from recycled frying oils comparable in quality to that of virgin vegetable oil biodiesel: both are composed of methyl esters of fatty acids and, therefore, they have very similar properties and potential in reducing pollutant emissions from the engine. Anyhow there is an imperative need to improve the existing biodiesel production methods from both economic and environmental viewpoints, and to investigate alternative and innovative production processes. The identification of some key parameters (acid value and FFA content, moisture content, viscosity, and fatty acid profile of the used oil) is a prerequisite for determining the viability of the vegetable oil transesterification process and, therefore, is essential for identifying the right processes to achieve the best results with respect to yield and purity of the produced biodiesel. In most cases, a simple pretreatment (removal by filtration of solid particles and esterification process to reduce the content of FFAs) is enough for subsequent transesterification. These results are expected to encourage the public and private sectors to improve the collection and recycling of used cooking oil to produce biodiesel. Using this fuel would not only avoid generating more waste, but also would provide a more eco-friendly fuel which cuts in half the greenhouse gases when compared to standard diesel fuels (Refaat, 2012).

First-generation bioethanol can be derived from renewable sources of virgin feedstock; typically starch and sugar crops such as corn, wheat, or sugarcane. Indeed most of the feedstocks used for first generation biofuel production are food crops. For this reason biofuel expansion can compete with food production directly (food crops diverted for biofuel production) and indirectly (competition for land and agricultural labor) (Gasparatos et al., 2011). These barriers can be partly overcome by the utilization of lignocellulosic materials for the production of the so-called second-generation bioethanol. One potential advantage for cellulosic ethanol technologies is that they can avoid direct competition for crops used in the food supply chain as it is not edible but this option should be limited to cases where actual and sustainable surplus of crops occurs or where crop wastes and wood wastes are available as a feedstock. (Timilsina and Shrestha, 2011; Pirozzi et al., 2012; Refaat, 2012). Cellulosic ethanol has a number of potential benefits over corn grain ethanol. Cellulosic ethanol is projected to be much more cost-effective, environmentally beneficial, and have a greater energy output to input ratio than grain ethanol. Cellulosic ethanol production, in particular, can result in a fuel with a net energy yield that is close to CO_2 neutral (Refaat, 2012). Although the cost of biomass is low, releasing fermentable sugars from these materials remains challenging.

Butanol can be obtained from food waste by fermentation processes using *Clostridium acetobutylicum* bacteria. This organism has a number of unique properties, including the ability to use variety of starchy substances and to produce much better yields of acetone and butanol than did Fernbach's original culture (Stoeberl et al., 2011). Butanol as fuel or blending component has some advantages compared to ethanol, for example a lower vapour pressure, improved combustion efficiency, higher energy density and it can be dissolved with vegetable oils in any ratio reducing their viscosity. Data on butanol production indicate that a potential of 0.3 g of butanol from 1 g carbohydrates from waste whey being a substrate characterised by high lactose content (Stoeberl et al., 2011).

Anaerobic digestion for biogas production (methane rich gas) is a well-established technology perfectly suitable for food waste management. In this framework, interest for anaerobic digestion (AD) has been continuously growing in the last decades, being more and more frequently promoted by national programmes for energy production from renewable resources. AD processes are also considered to be the best option for the biological production of hydrogen, the latter being recognized as one of the most interesting and promising biofuels (Guo et al., 2010; Ozkan et al., 2010; De Gioannis et al., 2013).

Numerous investigators demonstrated that if fermentation of biodegradable organic substrates is appropriately operated in a two-staged mode, separation of the acidogenic and methanogenic phases can be accomplished (i.a. De Giannis et al., 2013). Production of hydrogen by means of anaerobic digestion processes, combined with methane production (Kapdan and Kargi, 2006) may therefore represent an interesting solution for synergising sustainable management of food waste with renewable energy production.

A number of potentially suitable residual substrates have been evaluated for biohydrogen generation potential through dark fermentation. Among these, food waste may represent relatively inexpensive and suitable sources of biodegradable organic matter for H_2 production, mainly due to their high carbohydrate content and wide availability (i.a. De Gioannis et al., 2013).

1.3.4.2 BIOPRODUCTS PRODUCTION

The current frontier in the bioprocessing of organic materials lies in the biorefinery concept where organic waste is considered as a feedstock for the biological production of high value commodities. There is particular interest in the production of metabolites as renewable and biodegradable substitutes for petrochemical products. These metabolites include: lactate for the production of polylactate, a plastic constituent; polyhydroxyalkanotes, particularly polyhydroxybutyrate, which are natural storage polymer of many bacterial species with properties similar to polyethylene and polypropylene and harvestable from mixed cultures fed with organic wastes; succinate, a valuable and flexible precursor for pharmaceutical, plastic and detergent production, fermentable from carbohydrate rich wastes by selected bacterial species (Clarke and Alibardi, 2010).

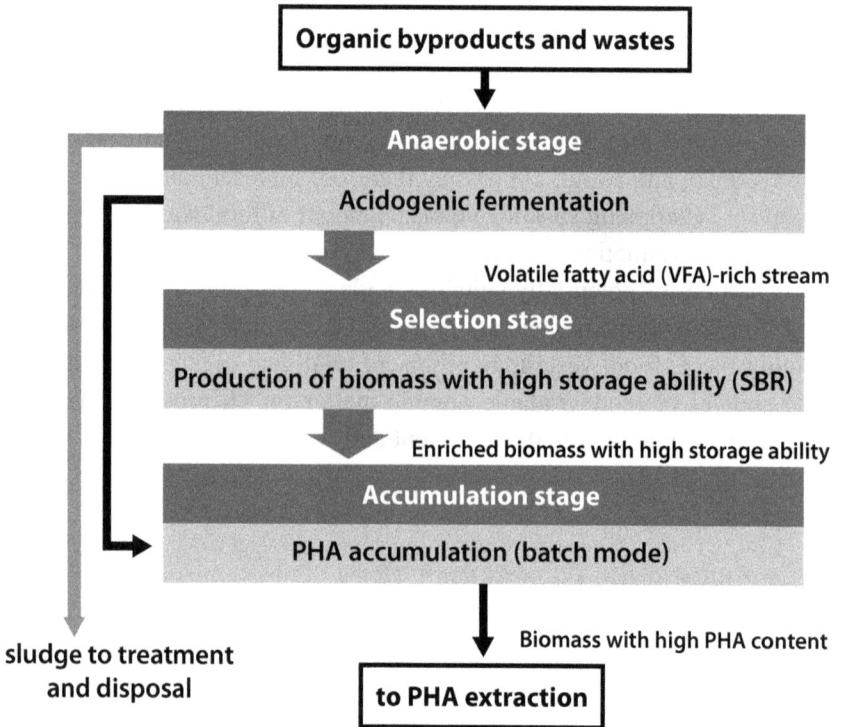

FIGURE 2: Schematic representation of a three-stage sequential process for Polyhydroxyalkanoate (PHA) production from waste/surplus-based feedstocks by mixed cultures using sequencing batch reactor (SBR) to carry out culture selection (from Reis et al., 2011).

The challenge of finite fossil resources has been addressed by academic and industrial research with the development of polymers based on renewable resources (besides the previously mentioned biofuels). The corresponding monomers are accessible either through fermentation of carbohydrate feedstocks by microbes, often genetically modified, or by chemical processing of plant oils (Fuessl et al., 2012). As for the biofuel production from virgin feedstocks, there is much debate about manufacturing bioplastics from natural materials and whether they are a negative impact to human food supply. 2.65 kg of corn are required to produce 1 kg of polylactic acid (PLA), the most common commercially compostable plastic. Since 270 million tonnes of plastic are made every year, replacing conventional plastic with corn-derived polylactic acid would remove 715 million tonnes of corn from the world's food supply (Harris, 2013). In this context, the opportunity to use waste food for bio-plastics seems a well-working option.

At present, among the various types of starch-based biodegradable plastics (such as PLA and PVA-polyvinylacetate), polyhydroxyalkanoates (PHA) is one of the most promising.

Polyhydroxyalkanoates (PHAs) are linear polyesters of hydroxyacids (hydroxyalkanoate monomers) synthesized by a wide variety of bacteria through bacterial fermentation (Poirier, 2003; Reis et al., 2011). Its strength and toughness are good, it is completely resistant to moisture and it has very low oxygen permeability. Therefore PHA can be used for bottles and water resistant film production (Van Wegen et al., 1998; Poirier, 2003). The simplest type of PHA is polyhydroxybutyrate (PHB). Pure PHB possesses several undesirable mechanical properties, so the copolymer poly(3-hydroxybutyrate-co-3-hydroxyvalerate) (PHBV) is generally used. Commercially-produced PHBV are marketed under the name "BIOPOL" (Van Wegen et al., 1998).

PHAs represent a large group of bacterial polyesters, which can include over 150 different hydroxyacids. The majority of bacteria synthesizing PHAs can be broadly subdivided into two groups. One group produces short-chain-length PHAs (SCL-PHAs) with monomers ranging from 3 to 5 carbons in length, while a distinct group synthesizes medium-chain-length PHAs (MCL-PHAs) with monomers from 6 to 16 carbons (Poirier, 2003). PHAs accumulate in bacteria cytoplasm as a high molecular weight polymer forming intracellular granules of 0.2–0.7mm in diameter. Typically, PHAs accumulate to a significant proportion of the cell dry weight when bacteria are grown in a media that is limited in a nutrient essential for growth (typically nitrogen or phosphorus), but with an abundant supply of carbon (for example glucose). Under these conditions, bacteria convert the extracellular carbon into an intracellular storage form, namely PHA. When the limiting nutrient is resupplied, intracellular PHA is degraded and the resulting carbon is used for growth (Poirier, 2003; Reis et al., 2011).

The main limitation in using bacterial PHAs as a source of biodegradable polymers is their production cost. In particular the average cost is by far the most significant contributor to overall PHB price, approximately two and a quarter times greater than the capital cost of equipment (Van Wegen et al., 1998). Using agro-industrial food waste as substrate instead of virgin feedstock of refined sugar such as glucose, sucrose and corn steep liquor could, again, represent the turning point. Sugarcane and beet molasses, cheese whey effluents, plant oils, swine waste liquor, vegetable and fruit wastes, effluents of palm oil mill, olive oil mill, paper mill, pull mill and hydrolysates of starch (e.g., corn and tapioca), cellulose and hemicellulose are all excellent alternatives characterized by a high organic fraction (Reis et al., 2011).

A three stage biotechnological process was proposed by Reis and coworkers (Reis et al., 2011) demostrating good potential for PHA production from waste/surplus-based feedstocks using enriched mixed cultures. The basic concept, reported in Figure 2, was based on an initial acidogenic fermentation phase of the feedstock, a second phase of selection and production of PHA-storing bacterial biomass under dynamic feeding, and the last phase where PHA was accumulated in batch conditions.

It is important to underline the main role played by the initial acidogenic fermentation stage that is needed to overcome the weak point of the process represented by the fact that most waste and surplus feedstocks contain several organic compounds that are not equally suitable for PHA production. Carbohydrates in fact are not stored by mixed cultures as PHA, but rather as glycogen. Thus, the acidogenic fermentation is an essential stage to increase the potential of producing PHA by mixed cultures from surplus-based feedstocks. Carbohydrates and other compounds are, in this way, transformed into VFA which are readily convertible into PHA (Reis et al., 2011).

Interconnection of biotechnological processes for the co-production of bio-fuels and bio-products represents therefore a key strategy to maximise the food waste utilisation and the potential income of the entire bioprocess chain.

1.5 CONCLUSIONS

The development of sustainable solutions for food waste management represents one of the main challenges for the current society. Sustainable solutions will be capable of exploiting the precious resources represented by food waste to reach social, economical and environmental benefits.

Several possibilities are currently available: reuse to feed animals, production of biofuels, exploitation to obtain monomers to be used for plastic production.

The benefits of recycling food waste are clear: bio-energy recovery, lower GHG emissions and lower space requirements for landfilling.

Data and analysis to assess the potential of food waste prevention are still required. The actions should interest all the sectors of food supply chains, from producers to retailers and consumers.

REFERENCES

1. Beylot A., Villeneuve J., Bellenfant G. (2013). Life Cycle Assessment of landfill biogas management: Sensitivity to diffuse and combustion air emissions. Waste Manage. 33, 401-411.
2. Cossu R., (2009). From triangles to cycles. Waste Manage. 29, 2915–2917
3. Culver D. (2013). Feed the hungry, reduce food waste by promoting and publicizing the obscure Bill Emerson Good Samaritan Act of 1996. Evergreen Digest: A Journal of Progress for the Rest of Us.
4. De Gioannis G., Muntoni A., Polettini A., Pomi R. (2013). A review of dark fermentative hydrogen production from biodegradable municipal waste fractions. Waste Manage. 33, 1345-1361.
5. EC (2013). European Commission - Stop food waste. http://ec.europa.eu/food/food/sustainability/index_en.htm
6. EPA (2014). http://www.epa.gov/foodrecovery/

7. FAO (1983). Food loss prevention in perishable crops.
8. FAO (2014). Food and Agriculture Organisation of United Nations, http://fa.org/
9. Fuessl A., Yamamoto M. and Schneller A. (2012). Opportunities in bi-based building blocks for polycondensates and vinyl polymers., Elsevier.
10. Garcia N. (2011). Food Waste Recycled as Biofuel. Stanford University.
11. Garrone P., Melacini M., Perego A. (2014). Opening the black box of food waste reduction. Food policy, 46, 129-139.
12. Glanz R., Schneider F. (2009). Causes of food waste generation in household. Proceedings Sardinia 2009, Twelfth International Waste Management and Landfill Symposium. S. Margherita di Pula, Cagliari. CISA Publisher, Italy.
13. Graham-Rowe E., Jessop D.C., Sparks P. (2014) Identifying motivations and barriers to minimising household food waste. Resources, Conservation and Recycling, 84, 15-23.
14. Guo X.M., Trably E., Latrille E., Carrere H., Steyer J-P. (2010). Hydrogen production from agricultural waste by dark fermentation: a review. Int. J. Hydrogen Energy 35, 10660-10673.
15. Gustavson J., Cederberg C., Sonesson U., van Otterdijk R and Meybeck A. (2011). Global food losses and food waste. Extent, causes and prevention. Rome.
16. Harris W. (2013). How long does it take for plastics to biodegrade?. How Stuff Works.
17. Harrison J. (2004). U.S. Wastes Half of Food Produced. UA News, Arizona.
18. Kapdan I.K., Kargi F. (2006). Bio-hydrogen production from waste materials. Enzyme Microb. Technol. 38, 569-582.
19. Katajajuuri J.M., Silvennoinen K., Hartikainen H., Heikkila L., Reinikainen A. (2014). Food waste in the Finnish food chain. Journal of Cleaner Production, http://dx.doi.org/10.1016/j.jclepro.2013.12.057.
20. Kojima R. and Ishikawa M. (2013). Prevention and Recycling of Food Wastes in Japan: Policies and Achievements. Resilent cities, Kobe University, Japan.
21. Lipinski B, Hanson C., Lomax J., Kitinoja L., Waite R. and Searchinger T. (2013). Reducing food loss and waste. Installment 2 of Creating a Sustainable Food Future, World Resources Institute, Washington DC.
22. Manfredi S., Tonini D., Christensen T.H. (2010). Contribution of individual waste fractions to the environmental impacts from landfilling of municipal solid waste. Waste Manage. 30, 433-440.
23. Mena C., Terry L.A., Williams A., Ellram L. (2014). Causes of waste across multi-tier supply networks: Cases in the UK food sector. Int. J. Production Economics, http://dx.doi.org/10.1016/j.ijpe.2014.03.012.
24. Nelles M., Arena U., Bilitewski B. (2010). Thermal waste treatment – An essential component of a sustainable waste treatment system. Waste Manage. 30, 1159-1160.
25. Nigam P.S., Singh A. (2011). Production of liquid biofuels from renewable resources, Elsevier.
26. Ozkan L., Erguder T.H., Demirer G.N. (2010). Investigation of the effect of culture type on biological hydrogen production from sugar industry wastes. Waste Manage. 30, 792-798.

27. Papargyropoulou E., Lozano R., Steinberger J., Wright N., bin Ujang Z. (2014) The food waste hierarchy as a framework for the management of food surplus and food waste. Journal of Cleaner Production, doi: 10.1016/j.jclepro.2014.04.020.

28. Pirozzi D., Ausiello A. and Yousuf A. (2012). Exploitation of lignocellulosic materials for the production of II generation biodiesel. Proceesing Venice 2012, Fourth International Symposium on Energy from Biomass and Waste. Cini Foundation, Venice Italy. CISA Publisher, Italy.

29. Poirier Y. (2003). Biopolimers, Elsevier.

30. Quested T.E., Marsh E., Stunell D., Parry A.D. (2013). Paghetti soup: The complex world of food waste behaviours. Resource Conservation and Recycling, 79, 43-51.

31. Rafaat A. (2012). Biofuels from Waste Materials, Elsevier.

32. Reis M., Albuquerque M., Villano M., Majone M. (2011). Mixed Culture Processes for Polyhydroxyalkanoate Production from Agro-Industrial Surplus/Wastes as Feedstocks, Elsevier.

33. Schneider F., Lebersorger S. (2009) Household attitudes and behaviour towards wasting food - A case study. Proceedings Sardinia 2009, Twelfth International Waste Management and Landfill Symposium. S. Margherita di Pula, Cagliari. CISA Publisher, Italy.

34. Silvennoinen K., Katajajuuri J.M., Hartikainen H., Jalkanen L., Koivupuro H.K., Reinikainen A. (2012). Food waste volume and composition in the finnish supply chain: special focus on food service sector. Proceesing Venice 2012, Fourth International Symposium on Energy from Biomass and Waste. Cini Foundation, Venice Italy. CISA Publisher, Italy.

35. Smill V. (2004). Improving efficiency and reducing waste in our food system. Environmental Science, 1, 17-26.

36. Stoeberla M., Werkmeistera R., Faulstichb M., Russa W. (2011). Biobutanol from food wastes – fermentative production, use as biofuel an the influence on the emissions. Procedia Food Waste, 1, 1868-1974.

37. Stuart T. (2009). Waste. Uncovering the global food scandal. London: Penguin.

38. Timilsina G.R., Shrestha A. (2011). How much hope should we have for biofuels?, Elsevier.

39. Thomsen N.I., Milosevic N., Bjerg P.L. (2012). Application of a contaminant mass balance method at an old landfill to assess the impact on water resources. Waste Manage. 32, 2406-2417.

40. Van Wegen R. J., Ling Y.and Middelberg A. P. J. (1998). Industrial production of polyhydroxyalkanoates using escherichia coli: an economic analysis. Trans IChemE, 76, 417-426.

41. Venkat K. (2011). The anatomy of food waste. Environmental Leader, http://www.environmentalleader.com/2011/09/28/the-anatomy-of-food-waste/

CHAPTER 2

Effect of Increasing Total Solids Contents on Anaerobic Digestion of Food Waste under Mesophilic Conditions: Performance and Microbial Characteristics Analysis

JING YI, BIN DONG, JINGWEI JIN, AND XIAOHU DAI

2.1 INTRODUCTION

Food waste (FW), usually from residential, commercial establishments, institutional and industrial sources, is generated at an ever-increasing rate (higher than 10% every year) with the rapid population growth and rising living standards in China [1]. It seems to be a good idea to reuse this favorable feedstock for energy recovery and municipal solid waste (MSW) reduction because FW contains high moisture and biodegradable organics and accounts for 40–50% of the weight of MSW. Anaerobic digestion (AD) is the most attractive and cost-effective technology for treating sorted organic fraction of MSW, especially food wastes [2]. Various AD

Effect of Increasing Total Solids Contents on Anaerobic Digestion of Food Waste under Mesophilic Conditions: Performance and Microbial Characteristics Analysis. © *Yi J, Dong B, Jin J, Dai X.* PLoS ONE **9**,7 *(2014).* *http://journals.plos.org/plosone/article?id=10.1371/journal.pone.0102548.* Licensed under Creative Commons International License, http://creativecommons.org/licenses/by/4.0/.

processes have been widely developed in many countries for the treatment of FW.

So far, three main types of AD technologies have been developed according to the total solids (TS) content of feedstocks: conventional wet (≤10% TS), semi-dry (10–20% TS) and modern dry (≥20% TS) processes. Dry anaerobic digestion, so called "high-solids" technology, has become attractive and was applied widely because it requires smaller reactor volume, lower energy requirements for heating, less material handling, and so on [2]–[4]. The TS content of solid waste influences anaerobic digestion performance, especially biogas and methane production efficiency [5]. Previous reports have investigated that role of TS content on AD performance in order to determine conditions for optimum gas production. Abbassi-Guendouz et al., showed that the total methane production decreased with TS contents increasing from 10% to 25% in batch anaerobic digestion of cardboard under mesophilic conditions [6]. The results obtained by Duan et al., showed that high-solids system could reach much higher volumetric methane production rate compared with low-solids system at the same solid retention time (SRT) in mesophilic anaerobic reactors treating sewage sludge [3]. Forster-Carneiro et al., showed that the biogas and methane production decreased with the total solids contents increasing from 20% to 30% in dry batch anaerobic digestion of food waste [2].

Anaerobic digestion is a multi-stage biochemical process in which the complex organic materials undergo hydrolysis, acidogenesis, and methanogenesis in series and each metabolic stage is functioned by different types of microorganisms [4]. They are present in a mixed culture but differ in their nutritional and pH requirement, growth kinetics, and their ability to tolerate environment stresses [7]. Characterization of microbial community structures in anaerobic digesters has been attractive from the point of review of engineering because understanding of microbial behavior can provides valuable information to optimize fermentation process to favor efficient breakdown of wastes [8]. However, the available literature is mainly about performance and corresponding the structure and dynamic of microbial community in either thermophilic or mesophilic anaerobic digestion of food waste, or only simply about performance comparisons. The AD performances at steady state and the comprehensive characterizations of microbial community in anaerobic digestion of FW with different

TS contents (wet, semi-dry and dry) were not compared in parallel. In order to increase the efficiency of anaerobic digestion of FW, it is necessary to understand the role of the TS contents on the behavior of the microbial community structure involved in the anaerobic digestion of degradation from wet to dry technology.

Recently, various molecular microbial ecology tools have been applied in numerous studies to analyze microbial communities in different anaerobic digesters and their influences on the efficiency and stability of AD processes [9]. Pyrosequencing, as a next generation sequencing technology, has gained increasing attention as a novel tool for studying the microbial diversity [4]. Recently, this technology has been widely and successfully used to characterize the microbial community structures in various environmental samples, such as source water [10], membrane filtration systems [11], soil [12]. Meanwhile, the microbial community structures were compared by this technology in anaerobic digestion of food waste at different organic loading rates (OLRs) [4].

Hence, the aim of this study was to conduct a comprehensive comparison of the microbial community structure using 454 high throughput pyrosequencing technology and related these microbial findings to their respective performances of mesophilic anaerobic digesters treating FW with different TS contents ranging from 5%–20%. It was expected that the reported work herein will reveal the role of the TS content on the behavior of the microbial community structure to increasing TS contents and hence to effective guide high solids anaerobic digestion of FW and to optimize the operational conditions for high anaerobic digestion efficiency.

2.2 MATERIALS AND METHODS

2.2.1 SUBSTRATES AND INOCULUMS

FW used in this study was collected every 30 days from a dining room at Tongji University in Shanghai. After removing bones, shells, and other indigestible materials, the FW was finely smashed using an electrical crusher and sufficient mixed and stored at 4°C. The TS of the FW ranged from 26% to 28% (w/w) and volatile solid (VS) accounted for 92%–95% of

TS. The mesophilic seed sludge was obtained from a full-scale anaerobic digester at Bailonggang municipal wastewater treatment plant (WWTP) (Shanghai, China). It had TS of 4.1% (w/w) and VS of 52.3% of TS. The main characteristics (average data plus standard deviations in duplicate tests) of substrates and inoculums are listed in Table 1. The collected FW was heated to 35°C before daily feeding.

TABLE 1: Characteristics of the substrates and inoculums.

Parameters	FW[a] 1 (days 1–30)	FW 2 (days 31–60)	FW 3 (days 61–90)	FW 4 (days 91–120)	Inoculums
TS[b] (%, w/w)	26.5 ± 0.6	27. 8 ± 1.1	27.3 ± 1.2	26.8 ± 1.2	4.1 ± 0.1
VS[c]/TS (%)	94.7 ± 3.9	92.2 ± 3.7	93.4 ± 4.6	93.9 ± 4.2	52.3 ± 2.4
pH	4.72 ± 0.21	4.64 ± 0.11	4.79 ± 0.24	4.87 ± 0.23	7.9 ± 0.3
C/N (w/w)	13.4 ± 0.6	14.2 ± 0.7	13.9 ± 0.4	13.6 ± 0.6	–
TAN[d] (mg/L)	538 ± 24	546 ± 19	534 ± 25	543 ± 19	299 ± 13

–Not determined. [a]FW: food waste. [b]TS: total solids. [c]VS: volatile solids. [d]TAN: total ammonia nitrogen

2.2.2 REACTORS AND OPERATION

Three identical reactors (numbered R1, R1 and R3), with liquid working volume of 6.0 L, were equipped with helix-type stirrers to provide sufficient mixing for substrates. The rotation speed was set at a rate of 60 rpm (rotations per minute) with 9 min stirring and 1 min break, continuously. Daily feeding was carried out by pushing semi-fluid substrate through the feeding piston. Since the digestate of FW in each reactor was completely fluid, daily draw-off was easily carried out by opening the discharge valve.

On the first day of the experiments, 6.0 L seed sludge was added to each reactor, which was operated semi-continuously (once-a-day draw-off and feeding) under single phase mesophilic conditions (35°C). The reactors were purged with N_2 for 10 min in order to provide anaerobic conditions. During the start-up period, the OLR was increased stepwise with high-solids FW before the TS content of the substrate in each reactor did not reach its designed TS level. Once the TS of the substrate in each reactor approached its designed level, the feeding FW was diluted to its designed TS level (5%, 15% and 20%, respectively) with de-ionized water before feeding. Each reactor was operated for five SRTs at 20 days SRT. For a full understading of the microbial community structures in anaerobic fermentation reactors with different TS contents, the anaerobically digested FW samples were taken on Day 100 when the systems could be deemed to have reached their steady state operation (determined by constant methane yield and VS reduction) after running for more than 3 months. The fermentation substrate samples in the reactors were taken every three days during the operation period of the fifth SRT for reactor performance analysis.

2.2.3 DNA EXTRACTION, PCR AND PYROSEQUENCING

To analyze the bacterial and archaeal communities in mesophilic anaerobic digesters with different feeding TS levels, 0.5 g of sample in reactor operated for 100 d was used for DNA extraction using a Fast DNA Spin Kit (QBIOgene, Carlsbad, CA, USA) following the manufacturer's instructions. For each sample, two independent PCR reactions were conducted using the primer pairs of 27F (5'-AGAGTTTGATCCTGGCT-CAG-3') and 533R (5'-TTACCGCGGCTGCTGGCAC-3') for bacteria and 344F (5'-ACGGGGYGCAGCAGGCGCGA-3') and 915R (5'-GT-GCTCCCCCGCCAATTCCT-3') for archaea [4]. To achieve the sample multiplexing during pyrosequencing, barcodes were incorporated in the 5'end of reverse primers 553R and 915R. All PCR reactions were carried out in a 25 uL mixture containing 0.5 uL of each primer at 30 mmolL^{-1}, 1.5 uL of template DNA (10 ng), and 22.5 uL of Platinum PCR SuperMix (Invitrogen, Shanghai, China). The PCR amplification program contained an initial denature at 95°C for 5 min, followed by 25 cycles of denaturing at 95°C for 30 s, annealing at 55°C for 30 s, and extension at 72°C for 30 s, followed by a final extension at 72°C for 5 min. The thermal cycling for archaea was similar to that for bacteria except that the annealing temperature was 57°C. After amplification, the PCR products were purified and quantified, and an equal amount of the PCR product was combined in a single tube to be run on a Roche GS FLX 454 Pyrosequencing machine at Majorbio Bio-Pharm Technology Co., Ltd., Shanghai, China.

2.2.4 ANALYSIS OF PYROSEQUENCING-DERIVED DATA

After sequencing completed, all sequence reads were quality checked using Mothur software [13]. Raw sequence reads were filtered before subsequence analyses to minimize the effect of random sequencing errors. The sequence reads that did not contain the correct primer sequence after the initial quality check (primer sequences were subsequently removed), were shorter than 200 bp, contained one or more ambiguous base(s), or checked as chimeric artifact were eliminated. Finally, the high-quality sequences after filtering were assigned to samples according to barcodes. Sequences were aligned in according with SILVA alignment [14]. Mothur was also used to conduct rarefaction curve, abundance base coverage estimator (ACE), richness (Chao), Shannon diversity, Simpson diversity indices and Good's coverage analysis, assign sequences to operational taxonomic units (OTUs, 97% similarity) using furthest neighbor approach. For taxonomy-based analysis, the SILVA database project (http://www.arb-silva. de) was used as a repository for aligned rRNA sequences. The sequences have been deposited into the NCBI short read archive (SRA) under the accession number SRX484115 for bacteria and SRX485028 for archaea.

2.2.5 ANALYTICAL METHODS

Volumes of produced biogas were measured by wet gas meters every day. The methane content of the biogas was measured by a gas chromatograph (GC) (Agilent Technologies 6890N, CA, USA) with a thermal conductivity detector equipped with Hayseq Q mesh and Molsieve 5A columns. For the analysis of volatile fatty acid (VFA), the fermentation mixtures withdrawn from digesters were centrifuged at 10, 000 ×g for 10 min, and then the supernatants were immediately filtered through 0.45 um cellulose nitrate membrane fiber paper. The filtrate was collected in a 1.5 ml gas chromatography (GC) vial and acidified by formic acid to adjust the pH to approximately 2.0, and then analyzed using a gas chromatograph (GC, Agilent 7820) with a flame ionization detector (FID) and equipped with a 52 CB column (30 m×0.32 mm×0.25 mm). The concentration of total VFA was calculated as the sum of the measured acetic, propionic, n-butyric iso-butyric, n-valeric, and iso-valeric acids. Metrohm 774 pH-meter was used in all pH measurements. The TS, VS, total alkalinity (TA) and total ammonia-nitrogen (TAN) were measured according to Standard Methods [15]. Free ammonia-nitrogen (FAN) was calculated in the same way as described by Østergaard [16]. The degradation or removal level based on VS (i.e., VS reduction) was calculated by the same formula as reported previously [3]. All experimental analyses were performed in triplicate. The data on performances of each reactor were expressed as mean±standard deviation of the samples.

2.3 RESULTS AND DISCUSSION

2.3.1 EFFECT OF TS CONTENT ON ANAEROBIC DIGESTION PERFORMANCE

Table 2 summarizes the values of the main parameters indicating system stability (pH, VFA, TA) and potential inhibitory chemicals (TAN and FAN) for three reactors operated at different TS contents, and the performance data were the average values of the last five samples during the operation period of the fifth SRT after the system reached steady state (determined by constant methane yield and VS reduction).

For each semi-continuously experiment with a good anaerobic digestion performance (between 5% and 20% TS), there was no accumulation of VFA and low pH. The concentration values of VFA showed increasing trend with increasing TS contents. Under mesophilic semi-dry anaerobic digestion of sorted organic fraction of municipal solid waste (OFMSW), Li et al., also observed an increasing trend of the VFA concentrations with TS contents increasing (for TS contents of 11.0%, 13.5% and 16.0%), the maximum VFA value was 4.2 g L^{-1}, 6.8 g L^{-1} and 22.4 g L^{-1}, respectively [17]. In this study, higher VFA concentrations were obtained in the reactors with higher TS contents, which could be explained by the fact that more organic matter was hydrolyzed and transformed to VFA in the reactors. High VFA levels and almost steady VS reduction (Table 3) in reactors indicated that the acidogenic activity was not influenced significantly. In addition, the reactor stability was maintained and the digestion occurred normally because a constant pH was maintained for each reactor. The average pH value was about 7.39, 7.68 and 7.82 at 5%, 15% and 20% TS, respectively. These pH values were within the permissible range for AD 6.5–8.5 but not with the optimal range 6.8–7.4 [18]. As we all know, the increase of VFA concentration contributes to the decrease of pH. However, low pH value was not observed in R3 in which the VFA concentration was highest. It could be explained by the fact that high buffering capacity was observed in high-solids anaerobic system at TS 20%, for which the total alkalinity value of 13.8 g $CaCO_3/L$ was detected.

TABLE 2: Summary of performance parameters on system stability and inhibition in three reactors.

reactor	SRT[a]	OLR[b] (Kg VS m^{-3} d^{-1})	pH	TA[c] (g/L)	TAN[d] (g/L)	FAN[e] (mg/L)	VFA[f] (g/L) Total	Acetic
R1 (5%)	20	2.35	7.39 ± 0.08	3.8 ± 0.1	0.40 ± 0.01	11 ± 0.4	0.12 ± 0.01	0.11 ± 0.01
R2 (15%)	20	7.01	7.68 ± 0.06	10.9 ± 0.3	1.31 ± 0.15	66 ± 2.5	0.53 ± 0.02	0.43 ± 0.01
R3 (20%)	20	9.41	7.82 ± 0.09	13.8 ± 0.2	1.92 ± 0.04	163 ± 8.0	0.94 ± 0.01	0.64 ± 0.02

[a]SRT: solid retention time. [b]OLR: organic loading rate. [c]TA: total alkalinity. [d]TAN: total ammonia nitrogen. [e]FAN: free ammonia nitrogen. [f]VFA: volatile fatty acid.

It was known that ammonia nitrogen concentration (especially free ammonia concentration) was an important factor influencing the stability of anaerobic digestion system. The TAN and FAN concentrations in three reactors at steady state were also observed. They showed a similar trend to that of above parameters with increasing TS contents. However, the maximum FAN value was just 163 mg/L. It has been reported that the FAN at concentrations above 200–1100 mg/L can inhibit the anaerobic system [19]. Therefore, the effect of FAN concentration on the system stability was probably negligible for the three reactors with TS contents ranged from 5% to 20%.

Biogas generation and methane efficiency of different reactors are shown in Table 3. The average daily cumulative biogas (based on added VS) of R1-5%, R2-15% and R3-20% accounted to 700, 760 and 870 ml and 370, 410 and 480 ml methane content, respectively. Hence, both of biogas production and methane content showed increasing trend with increasing TS contents. This result was in contrast with a previous work [2], in which the reactors with smaller TS contents showed higher biogas production and methane percentage in the batch anaerobic digestion of FW. It was suggested that the increasing of feeding TS contents lower than 20% has positive effect on the methane production. A maximum methane content of 55.1% in R3 agreed with the previous study on anaerobic digestion of FW [1], but was lower than in another reference [20], which was probably due to the differences in substrate composition. In addition, it could also be observed that reactors with higher TS contents showed higher volumetric biogas and methane production rate. It is well known that FW is a high degradable substrate for anaerobic digestion [21]. For reactors R1-R3 at a fixed 20 days SRT, increased feeding TS content of FW meant higher applied OLR and larger proportion of easily degradable substrate for microorganisms, which results in higher volumetric biogas yield and methane production rate. As showed in Table 3, higher VS reduction was observed in the anaerobic digesters with higher TS contents. The reasons for this important result obtained were investigated from the microbiology aspect in the following chapters. The specific biogas and methane product rate based on removed VS increased slightly. The highest specific biogas production rate determined on removed VS was 1.01 L gVS^{-1} removed in R3, which was higher than corresponding data obtained in a previous study [1].

TABLE 3: Performance parameters of three reactors with different total solids contents.

Reactor	SRT (d)	OLR (Kg VS m^{-3} d^{-1})	Y$_{biogas}$[a] (LBio-gas gVS$^{-1}_{added}$)	CH$_4$ (%)	Y$_{methane}$[b] (L CH$_4$ g VS$^{-1}_{added}$)	VS$_r$[c] (%)	SBP[d] (L Biogas gVS$^{-1}_{removed}$)	SMP[e] (LCH$_4$ gVS$^{-1}_{removed}$)	BP[f] (Biogas L^{-1} d^{-1})	MP[g] (LCH$_4$ L^{-1} d^{-1})
R1 (5%)	20	2.35	0.70 ± 0.02	52.5 ± 2.1	0.37 ± 0.01	80.1 ± 2.4	0.88 ± 0.02	0.46 ± 0.01	1.65 ± 0.06	0.87 ± 0.03
R2 (15%)	20	7.01	0.76 ± 0.01	54.2 ± 2.7	0.41 ± 0.01	82.4 ± 2.2	0.92 ± 0.05	0.50 ± 0.01	5.36 ± 0.2	2.90 ± 0.07
R3 (20%)	20	9.41	0.87 ± 0.02	55.1 ± 2.6	0.48 ± 0.01	85.6 ± 2.6	1.01 ± 0.04	0.56 ± 0.02	8.21 ± 0.34	4.52 ± 0.05

[a]Y_{biogas}: biogas yield. [b]$Y_{methane}$: methane yield. [c]VS_r: VS reduction. [d]SBP: specific biogas production rate based on removed VS. [e]SMP: specific methane production rate based on removed VS. [f]BP: volumetric biogas production rate. [g]MP: volumetric methane production rate.

2.3.2 OVERALL ANALYSIS OF PYROSEQUENCING

The latest developed 454 high-throughput pyrosequencing that can generate huge amounts of DNA reads is widely employed to investigate the bacterial and archaeal community structures and dynamics in various environmental samples. To investigate the compositions of microbial populations involved in the fermentative reactors with different TS contents, a total of 9571, 7769 and 5598 trimmed bacterial 16S rRNA gene sequences and 5245, 4654 and 4432 trimmed archaeal 16S rRNA gene sequences were recovered from samples R1, R2 and R3 (Table S1), respectively. The sequences were grouped into OTUs at a distance level of 3% to estimate the phylogenetic diversities of microbial communities. The OTUs number identified by bacterial sequences in R1 was the largest among three samples. The bacterial community richness levels can also reflected using ACE, Chao, Shannon and Simpson diversity indices (Table S1), which also revealed that the R1-5% had the highest bacterial diversity among three samples. However, the number of archaeal OTUs in R2 was the largest. The rarefaction curves of three samples generated at 3% cutoff for bacterial and archaeal communities are shown in the Figure S1 (Supporting information), demonstrating clearly that the bacterial community richness of R1 and the archaeal community richness of R2 was the highest among these samples, respectively. However, none of the curves approached a plateau, suggesting that this sequencing depth was still not enough to cover the whole microbial diversity and further sequencing would have resulted in more OTUs for each sample. Pyrosequencing analysis of environmental samples can obtain much more sequences and OTUs than conventional cloning and sequencing methods [11], [12]. The bacterial (or archaeal) PCR amplicons from anaerobic digester were grouped into only 238–514 (or 8–26) OTUs according to the clone library in a previous publication [22]. To the authors's knowledge, this was the first study using pyrosequencing technology to characterize the microbial communities in anaerobic digesters with different TS contents. It can be found that compared with traditional clone library, 454 high-throughput pyrosequencing could be a powerful tool to elucidate the microbial community structures and diversities in anaerobic reactors treating food waste with different TS contents.

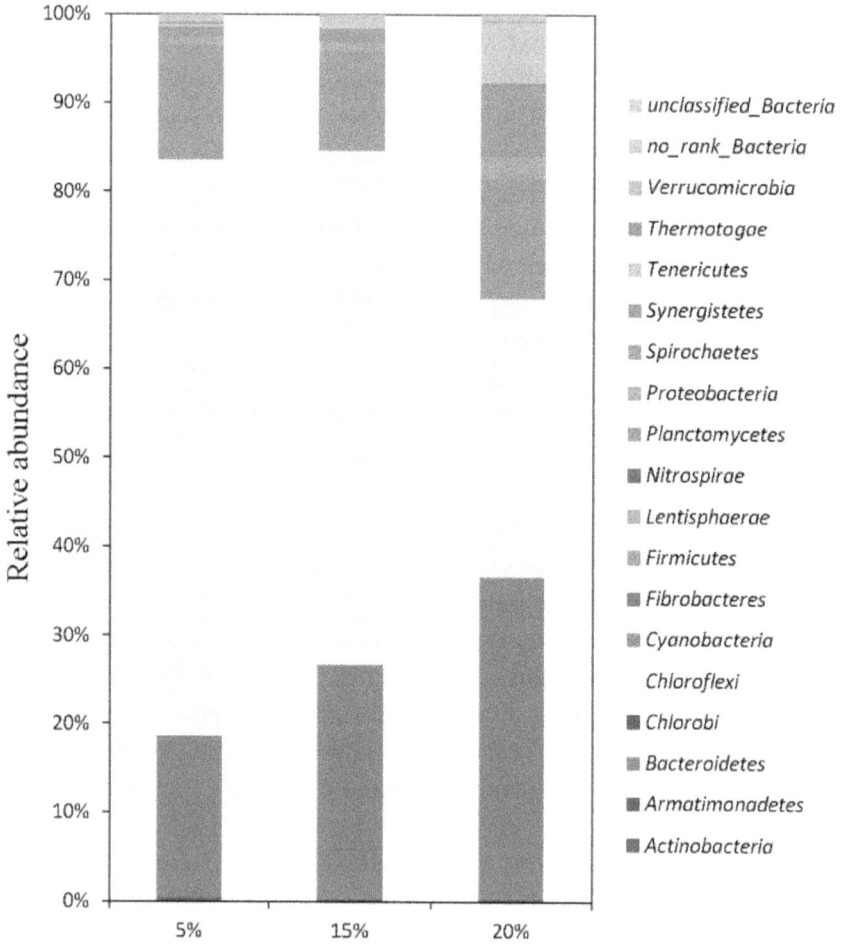

FIGURE 1: Taxonomic compositions of bacterial communities at phyla level in each sample retrieved from pyrosequencing.

TABLE 4: Taxonomic composition of bacterial communities at the genus level for the sequences retrieved from each sample.

Phylum	Genus	Relative abundance		
		5%	15%	20%
	Bacteroidales	0.43%	0.40%	1.54%
	Bacteriodes	0.54%	0.27%	0.82%
	Barnesiella	2.83%	0.08%	0.00%
	Marinilabiaceae	0.37%	0.22%	0.61%
Bacteriodetes	*Parabacteroides*	0.18%	0.06%	0.34%
	Petrimonas	0.45%	0.37%	1.11%
	Proteiniphilum	1.15%	2.59%	4.29%
	Rikenellaceae	11.16%	21.70%	26.58%
	Sphingobacteriales	0.48%	0.17%	0.52%
Chloroflexi	*Anaerolineaceae*	64.99%	58.03%	31.37%
	Anaerobranca	0.11%	0.36%	1.43%
	Christensenellaceae	0.30%	0.14%	0.18%
	Clostridiales	1.73%	4.00%	3.88%
	Erysipelotrichaceae	7.17%	1.85%	2.41%
Firmicutes	*Fastidiosipila*	0.30%	0.79%	1.09%
	Gelria	0.85%	0.85%	0.45%
	Lachnospiraceae	0.13%	0.72%	0.39%
	Lutispora	0.09%	0.06%	0.36%
	Ruminococcaceae	1.11%	0.75%	0.86%
Proteobacteria	*Novosphingobium*	0.21%	0.14%	0.77%
	Rhizobiales	0.07%	0.06%	0.36%
Spirochaetes	*Spirochaeta*	0.28%	0.26%	0.73%
	Spirochaetes	0.46%	1.08%	6.98%
Tenericutes	*Acholeplasma*	0.40%	1.12%	6.75%
	Minor group	4.21%	3.93%	6.20%

2.3.3 EFFECT OF TOTAL SOLIDS CONTENT ON FUNCTIONAL BACTERIAL POPULATIONS INVOLVED IN FOOD WASTE HYDROLYSIS AND ACIDIFICATION

Large numbers of bacterial populations are involved in the hydrolysis and acidification processes of anaerobic fermentation for food waste. The distribution of sequences at the phylum level in each sample is shown in Figure 1. There are seven phyla with relative abundance of higher than 0.5% in at least one sample. From the phylum assignment results, it can be seen that most bacterial sequences in the anaerobic digester treating food waste were distributed among three major phyla: *Chloroflexi, Bacteroidetes* and *Firmicutes*, the total relative abundances of them accounted for 96.13%, 95.61% and 81.35% in R1, R2 and R3, respectively, along with other phyla at minor predominance. Similarly, *Bacteroidetes, Firmicutes, Chloroflexi, Synergistetes,* and *Actinobacteria* were reported to the major populations at phylum level in the mesophilic anaerobic digester treating food waste [4]. The dominance of *Bacteroidetes, Firmicutes* and *Chloroflexi* was also found in other previous studies [7], [23]. In addition, R3 with 20% feeding TS content had high relative abundance of *Spirochaetes* (8.09%), *Tenericutes* (6.86%) and *Proteobacteria* (2.16%).

Although most bacteria in reactors were affiliated to these dominant phyla, the relative abundances of these phyla in each reactor were different and each digester had its own characteristic bacterial community composition. The proportion of phylum *Chloroflexi* in each reactor was the highest in this study. This was in good accordance with previous reports that *Choroflexi* populations were abundant in anaerobic digesters, as determined by membrane hybridization [7], FISH [23] and 16S rRNA gene clone analysis [23], [24]. Rivière et al., also found large proportions (25–45%) of *Chloroflexi* sequences in municipal WWTP sludge samples [22]. An important trend is the small proportion of *Choroflexi* at the highest TS content: 31% for the 20% TS, compared to 58% with the 15% TS and 65% at the 5% TS. The proliferation of *Choroflexi* (formerly known as Green Nonsulfur Bacteria), a well known scavenger biomass-derived organic carbon such as soluble microbial products (SMP), supports a greater influence of difficult-to-biodegrade organic materials from the input substrates and from endogenous dacay of the anaerobic biomass [22], [25].

For R1-R3 at a fixed SRT, increased feeding TS of FW meant higher applied OLR and larger amount of easily degradable substrate per unit volume for microorganisms, which resulted in a smaller relative abundance of phylum *Choroflexi*.

On the other hand, the *Bacteroidetes* population was enriched in the reactors with higher TS contents (from 18.2% at the 5% TS to 26.40% at the 15% TS and 36.33% at the 20% TS). The phylum *Bacteroidetes* are proteolytic bacteria and were probably involved in the degradation of various proteins used for anaerobic digestion studies [22], [25]. The majority of proteolytic microorganisms are able to metabolize amino acids to produce VFA such as acetate, propionate and succinate and NH_3 [22]. Interestingly, their selective enrichment at high TS contents seems to be in consistent with the observation of high protein-input rate and VFA production in the reactors with higher TS contents (Table 2). This result indicated the importance of the *Bacteroidetes* performing protein hydrolysis. However, the changing trend of relative abundance of the phylum *Firmicutes* was not obvious with increasing TS contents. The average value of *Firmicutes* proportion was 12% in three reactors. *Firmicutes* are well-known to be acetogenic and syntrophic bacteria that can degrade VFA, such as butyrate and its analogs. The prevalence of organisms belonging to *Firmicutes* suggested that these products are readily available due to the prior fermentation of these simple VFA and played a critical role in anaerobic digestion of FW, especially on the production of acetic acid, an essential step for methane production by acetoclastic methanogenic microorganisms. In addition, the relative abundances of other phyla including *Proteobacteria, Spirochaetes* and *Tenericutes* obviously increased with the feeding TS contents increasing. It has been suggested that they might play important roles in the degradation of FW. *Proteobacteria* are also involved in the first step of the degradation of organic wastes and they are important consumers of propionate, butyrate, and acetate [23]. *Spirochaetes* are reported to ferment carbohydrates or amino acids into, mainly, acetate, H_2 and CO_2 [8] and *Tenericutes* was found to be related with lignin utilization [26].

In order to further compare the difference of bacterial communities in anaerobic digesters with different feeding TS contents, it is preferable to deconstruct the sequencing date at the subdivision level. Therefore, the relative abundance of each genus in three samples was calculated. The sequence distributions at genus level in each sample are shown in Table 4. A total of 17 genera were detected among which 7 genera with relative abundance of higher than 0.5% in at least one sample were screened as the abundant genera. Other genera were grouped into the minors. As mentioned in the previous section, lower proportions of population from the phylum *Choroflexi* were markedly detected in the reactors with higher TS contents. All sequences classified to phylum *Choroflexi* in three reactors were assigned to genus *Anaerolineaceae* (Figure 1 and Table 4) and class *Anaerolineae* at class level (previous known as "subphylum I" [24]) (Table S2), and the relative abundance of genus *Anaerolineaceae* decreased with increasing TS contents. Because all the characterized species of the class *Anaerolineae* are anaerobic bacteria that decompose carbohydrates via fermentation [27], the genus *Anaerolineaceae* seemed to be involved in carbohydrate decomposition in anaerobic digestion of FW. Similarly, in the previous studies, it was found that all the *Choroflexi* sequences obtained from the up-flow anaerobic sludge blanket reactors treating various food-processing and high-strength organic wastewaters belong to the class *Anaerolineae* [27] and *Anaerolineaceae* group was dominant in phylum *Choroflexi* with its maximum proportion of 8.9% at the 58 days in mesophilic anaerobic digestion of FW [4].

Concerning *Bacteroidetes*, another very abundant phylum which increased with the increasing TS contents, the subdivisions at genus level were multiple and many genera were mainly presented in three anaerobic reactors. *Rikenellaceae* spp. and *Proteiniphilum* spp. were the mostly major genera within this dominant phylum and the changing trends of the relative abundances of these two genera were the same as that of the *Bacteroidetes*. *Rikenellaceae* spp. showed a remarkable proportion from 11% to 27%. The genus *Rikenellaceae* could utilize lactate as substrate in the fermentation processes, and acetate and propionate are the main end-products [28]. *Proteiniphilum*, a relatively new genus showed an unusual ability to grow well at 20–45°C and pH 6.0–9.7. The strains were proteolytic and yeast extract, peptone and l-arginine could be used as carbon and energy sources. Acetic acid and NH_3 were produced after utilizing these substrates [29]. The predominance of *Proteiniphilum* was also obtained in other anaerobic digesters by using a meta-analysis approach [9]. Other genera in this phylum with individual proportion higher or lower than 0.5% might also have played important roles in FW degradation. Regarding to *Firmicutes*, the generic distributions were also distinct with genera *Clostridiales* and *Erysipelotrichaceae* as the main groups in three anaerobic reactors (Table 4). The latter was especially notable in sample R1-5% with relative abundance of 7.13%. Moreover, the proportion of the reigning genera *Spirochaetes* within the abundant phylum *Spirochaetes* and *Acholeplasma* within the *Tenericutes* increased obviously with TS contents increasing. From the analyses made above, it can be seen that the changing patterns of main microbial population abundances were closely related to the performance variations with TS contents increasing, especially for VS reduction. The increasing degradation of organic matter to precursors for methanogenesis was jointly accomplished by the compatible collaborations of these microorganisms which played their respective roles in one of several trophic levels including hydrolysis, fermentation and acetogenesis.

2.3.4 EFFECT OF TOTAL SOLIDS CONTENT ON FUNCTIONAL ARCHAEAL POPULATIONS INVOLVED IN FOOD WASTE METHANOGENESIS

The diversities of archaeal populations in three anaerobic digesters were also revealed by high-throughput pyrosequencing target 16S rRNA gene segments. All species richness estimators including ACE, Chao Shannon and Simpson indices are shown in Table S1. The Good's coverage estimated at least 97% coverage at a similarity of 97%, indicating good coverage of archaeal community. Two hydrogen-utilizing methanogenic groups, *Methanobacteriales* and *Methanomicrobiales*, and acetoclastic methanogenic order *Methanosarcinales* were detected in three reactors. The sum relative abundances of these three methanogenic groups accounted for 99.64%, 99.19% and 99.62% of total archaeal sequences in R1, R2, R3, respectively. However, *Methanococcales* was not detected in any DNA samples in this study (Figure 2). This result was in accordance with previous work characterizing the microbial community shifts in anaerobic digestion of secondary sludge [30]. The relative abundances of sequencing data were also analyzed more specifically at genus level (Table. S3). It was showed that the phylogenetic diversity of methanogens was much lower than that of the bacterial community due partly to the inherent phylogenetic low diversity of methanogens.

As shown in Table S3, there was no large gap in terms of methanogens diversity and distinct discrimination in the taxonomic compositions at genus level. Most of methanogens were assigned to the genus *Methanosarcina* (accounting for 84.4%, 89.5% and 90.9% of total archaeal sequences in R1, R2, R3, respectively), indicating that acetoclastic methanogens played important roles in anaerobic digestion of FW and acetoclastic methanogenesis was the principal pathway of methane production. The low-solids anaerobic digester R1 was secondly dominated by hydrogenotrophic *Methanoculleus* while another hydrogenotrophic methanogens *Methanomicrobiales* was the second most detected group in anaerobic digesters R2 and R3.

Methanosarcina, a typical member of acetoclastic methanogens, have been often reported as the dominant methanogens in AD [31]. The ability of genus *Methanosarcina* having high growth rates and forming irregular cell clumps makes them more tolerant to changing in pH and high concentrations of toxic ionic agents [32]. The genus *Methanosarcina* produce methane from acetate, although some species are more versatile and can also utilize H_2/CO_2, methylated amines and methanol. In addition, *Methanosarcina* spp. are able to use both the acetoclastic and the hydrogenotroph methanogenesis pathways, making them more tolerant to specific inhibitors of the acetoclastic pathway compared to *Methanosaeta* spp. Therefore, anaerobic digester dominantly based on *Methanosaricna* spp. could potentially achieve stable methanogenesis [33], as their special morphological characteristics and flexibility in metabolism.

Besides, the changing patterns of the proportions of three major genera with TS contents increasing were different. The relative proportion of the genus *Methanosaricna* slightly increased from 84% to 90.9% with the TS content increased from 5% in R1 to 20% in R3. On the basis of stable operation, increased feeding TS contents of FW meant higher applied OLR and more VS for microorganisms, which resulted in higher VFA concentrations. In this study, it was observed in Table 2 that the acetate concentration increased with increasing TS contents. It is suggested that higher acetate concentrations would favor the growth of *Methanosarcina* [33].

Therefore, higher concentrations of VFA (especially acetate) and, by extension, at higher OLR caused by the anaerobic systems with higher TS contents induced the selective proliferation of *Methanosarcina*.

The relative abundance of genus *Methanoculleus* obvious decreased from 7.63 to 2.91% with increasing TS contents, indicating that hydrogenotrophic methanogenesis by *Methanoculleus* contributed less to the methane production in high-solids AD than it did in low-solids AD. It has been reported that *Methanoculleus* methanogens had been widely distributed with large proportion in various thermophilic ananerobic digesters [34] and their population ratio seems to be affected by HRT, OLR, or the concentration of VFA. In this study, similar result was obtained that the dominance of *Methanoculleus* declined in the mesophilic anaerobic digesters with TS content increasing resulting in the increase of OLR and the concentration of VFA. Summarily, the changing of microbial communities in mesophilic anaerobic digestion of FW was responsible for the different performances of the reactors with the increasing TS contents. The results obtained in this study expand our knowledge about the role of the TS content on the behavior of the microbial community structure involved in the anaerobic digestion degradation of solids, from low-solids to high-solids technology, and hence to provide valuable information to optimize fermentation process to favor efficient breakdown of food waste.

2.4 CONCLUSIONS

Three stable processes were achieved for AD of food waste with TS contents increasing from 5% to 20%. Better performances, mainly including VS reduction and methane yield and significant shifts in bacterial community, were obtained with the increasing TS contents. The relative abundance of phylum *Chloroflexi* decreased while other functional bacteria increased. The genus *Methanosarcina* absolutely dominated in archaeal communities in three reactors and the relative abundance of this group showed increasing trend with TS contents increasing. These results revealed the effect of the TS content on the performance parameters and the behavior of the microbial community involved in the AD of food waste from wet to dry technologies.

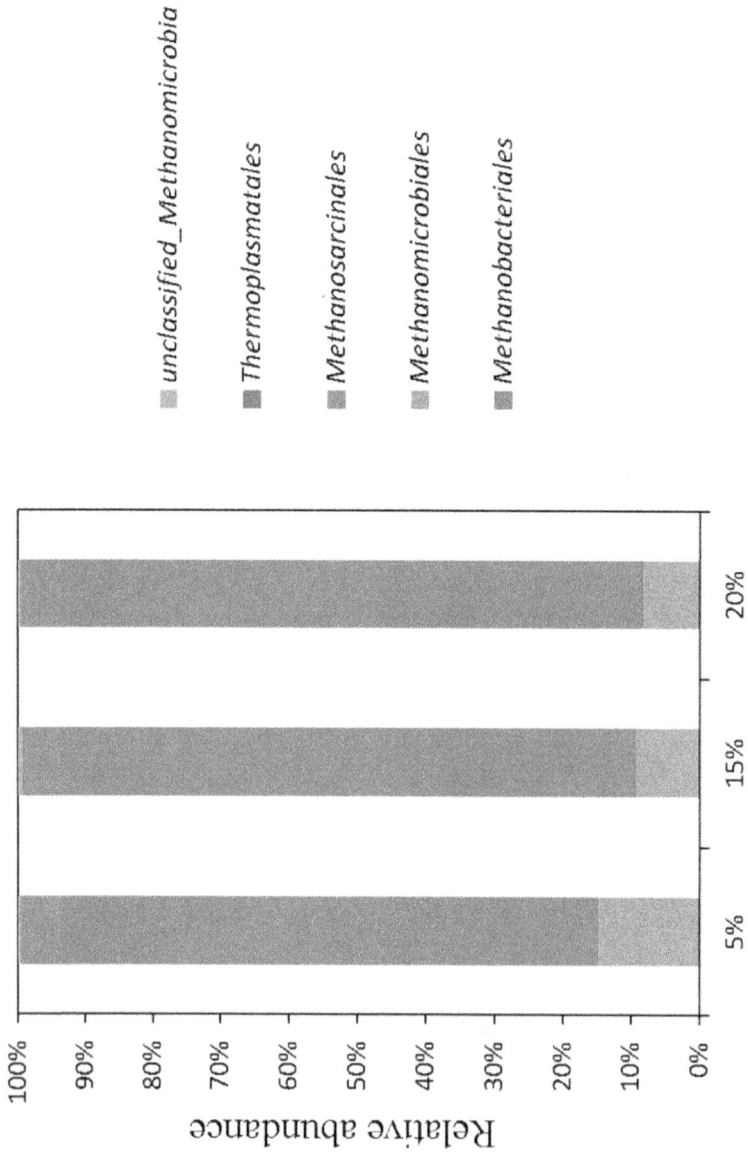

FIGURE 2: Taxonomic compositions of methanogens at order level in each sample retrieved from pyrosequencing.

REFERENCES

1. Dai X, Duan N, Dong B, Dai L (2013) High-solids anaerobic co-digestion of sewage sludge and food waste in comparison with mono digestions: Stability and performance. Waste Manage 33: 308–316. doi: 10.1016/j.wasman.2012.10.018

2. Forster-Carneiro T, Pérez M, Romero L (2008) Influence of total solid and inoculum contents on performance of anaerobic reactors treating food waste. Bioresour Technol 99: 6994–7002. doi: 10.1016/j.biortech.2008.01.018

3. Duan N, Dong B, Wu B, Dai X (2012) High-solid anaerobic digestion of sewage sludge under mesophilic conditions: feasibility study. Bioresour Technol 104: 150–156. doi: 10.1016/j.biortech.2011.10.090

4. Guo X, Wang C, Sun F, Zhu W, Wu W (2014) A comparison of microbial characteristics between the thermophilic and mesophilic anaerobic digesters exposed to elevated food waste loadings. Bioresour Technol 152: 420–428. doi: 10.1016/j.biortech.2013.11.012

5. Pavan P, Battistoni P, Mata-Alvarez J (2000) Performance of thermophilic semi-dry anaerobic digestion process changing the feed biodegradability. Water Sci Technol 41: 75–81.

6. Abbassi-Guendouz A, Brockmann D, Trably E, Dumas C, Delgenès JP, et al. (2012) Total solids content drives high solid anaerobic digestion via mass transfer limitation. Bioresour Technol 111: 55–61. doi: 10.1016/j.biortech.2012.01.174

7. Chouari R, Le PD, Daegelen P, Ginestet P, Weissenbach J, et al. (2005) Novel predominant archaeal and bacterial groups revealed by molecular analysis of an anaerobic sludge digester. Environ Microbiol 7: 1104–1115. doi: 10.1111/j.1462-2920.2005.00795.x

8. Fernández A, Huang S, Seston S, Xing J, Hickey R, et al. (1999) How stable is stable? Function versus community composition. Applied Environ Microb 65: 3697–3704.

9. Nelson MC, Morrison M, Yu Z (2011) A meta-analysis of the microbial diversity observed in anaerobic digesters. Bioresour Technol 102: 3730–3739. doi: 10.1016/j.biortech.2010.11.119

10. Pinto AJ, Xi C, Raskin L (2012) Bacterial community structure in the drinking water microbiome is governed by filtration processes. Environ Sci Technol 46: 8851–8859. doi: 10.1021/es302042t

11. Kwon S, Moon E, Kim TS, Hong S, Park HD (2011) Pyrosequencing demonstrated complex microbial communities in a membrane filtration system for a drinking water treatment plant. Microbes Environ 26: 149–155. doi: 10.1264/jsme2.me10205

12. Roesch L, Fulthorpe RR, Riva A, Casella G, Hadwin AK, et al. (2007) Pyrosequencing enumerates and contrasts soil microbial diversity. ISME J 1: 283–290. doi: 10.1038/ismej.2007.53

13. Schloss PD, Westcott SL, Ryabin T, Hall JR, Hartmann M, et al. (2009) Introducing mothur: open-source, platform-independent, community-supported software for describing and comparing microbial communities. Appl Environ Microb 75: 7537–7541. doi: 10.1128/aem.01541-09

14. Quast C, Pruesse E, Yilmaz P, Gerken J, Schweer T, et al. (2013) The SILVA ribosomal RNA gene database project: improved data processing and web-based tools. Nucleic acids res 41: 590–596. doi: 10.1093/nar/gks1219

15. APHA (American Public Health Association) (1995) Standard Methods for the Examination of Water and Wastewater, 19th ed, Washington, DC, USA.

16. Østergaard N (1985) Biogasproduktion i det thermofile temperaturinterval: Kemiteknik, Teknologisk Institut.

17. Li R, Chen S, Li X (2010) Biogas production from anaerobic co-digestion of food waste with dairy manure in a two-phase digestion system. Appl Environ Microb 160: 643–654. doi: 10.1007/s12010-009-8533-z

18. Malina J, Pohland JF, Frederick G (1992) Design of anaerobic processes for the treatment of industrial and municipal wastes: CRC Press 7.

19. Hansen KH, Angelidaki I, Ahring BK (1998) Anaerobic digestion of swine manure: inhibition by ammonia. Water Res 32: 5–12. doi: 10.1016/s0043-1354(97)00201-7

20. Lou XF, Nair J, Ho G (2012) Field performance of small scale anaerobic digesters treating food waste. Energy Sustain Dev 16: 509–514. doi: 10.1016/j.esd.2012.06.004

21. Heo NH, Park SC, Kang H (2004) Effects of mixture ratio and hydraulic retention time on single-stage anaerobic co-digestion of food waste and waste activated sludge. J Environ Sci Heal A 39: 1739–1756. doi: 10.1081/ese-120037874

22. Rivière D, Desvignes V, Pelletier E, Chaussonnerie S, Guermazi S, et al. (2009) Towards the definition of a core of microorganisms involved in anaerobic digestion of sludge. ISME J 3: 700–714. doi: 10.1038/ismej.2009.2

23. Ariesyady HD, Ito T, Okabe S (2007) Functional bacterial and archaeal community structures of major trophic groups in a full-scale anaerobic sludge digester. Water Res 4: 1554–1568. doi: 10.1016/j.watres.2006.12.036

24. Yamada T, Sekiguchi Y (2009) Cultivation of uncultured *Chloroflexi* subphyla: significance and ecophysiology of formerly uncultured *Chloroflexi* 'subphylum I' with natural and biotechnological relevance. Microbes Environ 24: 205–216. doi: 10.1264/jsme2.me09151s

25. Kindaichi T, Ito T, Okabe S (2004) Ecophysiological interaction between nitrifying bacteria and heterotrophic bacteria in autotrophic nitrifying biofilms as determined by microautoradiography-fluorescence in situ hybridization. Appl Environ Microb 70: 1641–1650. doi: 10.1128/aem.70.3.1641-1650.2004

26. Boucias DG, Cai Y, Sun Y, Lietze VU, Sen R, et al. (2013) The hindgut lumen prokaryotic microbiota of the termite Reticulitermes flavipes and its responses to dietary lignocellulose composition. Mol EcoL 22: 1836–1853. doi: 10.1111/mec.12230

27. Narihiro T, Terada T, Kikuchi K, Iguchi A, Ikeda M, et al. (2008) Comparative analysis of bacterial and archaeal communities in methanogenic sludge granules from upflow anaerobic sludge blanket reactors treating various food-processing, high-strength organic wastewaters. Microbes Environ 24: 88–98. doi: 10.1264/jsme2.me08561

28. Su Y, Li B, Zhu WY (2012) Fecal microbiota of piglets prefer utilizing dl-lactate mixture as compared to d-lactate and l-lactate in vitro. Anaerobe 19: 27–33. doi: 10.1016/j.anaerobe.2012.11.006

29. Chen S, Dong X (2005) Proteiniphilum acetatigenes gen nov, sp nov, from a UASB reactor treating brewery wastewater. Int J Syst Evol Micr 55: 2257–2261. doi: 10.1099/ijs.0.63807-0

30. Shin SG, Lee S, Lee C, Hwang K, Hwang S (2010) Qualitative and quantitative assessment of microbial community in batch anaerobic digestion of secondary sludge. Bioresour Technol 101: 9461–9470. doi: 10.1016/j.biortech.2010.07.081

31. Demirel B, Scherer P (2008) The roles of acetotrophic and hydrogenotrophic methanogens during anaerobic conversion of biomass to methane: a review. Rev Environ Sci Biotechnol 7: 173–190. doi: 10.1007/s11157-008-9131-1

32. Conklin A, Stensel HD, Ferguson J (2006) Growth kinetics and competition between *Methanosarcina* and Methanosaeta in mesophilic anaerobic digestion. Water Environ Res 78: 486–496. doi: 10.2175/106143006x95393

33. Vrieze JD, Hennebel T, Boon N, Verstraete W (2012) *Methanosarcina*: the rediscovered methanogen for heavy duty biomethanation. Bioresour Technol 112: 1–9. doi: 10.1016/j.biortech.2012.02.079

34. Bourque JS, Guiot S, Tartakovsky B (2008) Methane production in an UASB reactor operated under periodic mesophilic-thermohilic conditions. Biotechnol Bioeng 100: 1115–1121. doi: 10.1002/bit.21850

There are several supplemental files that are not available in this version of the article. To view this additional information, please use the citation on the first page of this chapter.

CHAPTER 3

Microbial Anaerobic Digestion (Bio-Digesters) as an Approach to the Decontamination of Animal Wastes in Pollution Control and the Generation of Renewable Energy

CHRISTY E. MANYI-LOH, SAMPSON N. MAMPHWELI, EDSON L. MEYER, ANTHONY I. OKOH, GOLDEN MAKAKA, AND MICHAEL SIMON

3.1 INTRODUCTION

Biomass encompasses materials derived from plants, animals, humans as well as their wastes. In addition, food processing, agro-industrial and industrial wastes are other sources of biomass wastes, so also is the microbial population metabolically active and cultivable plus metabolically active but non-cultivable cells existing within these wastes. Depending on the characteristics of these wastes, they can be converted into energy/and or fuel by combustion, gasification, co-firing with other fuels and ultimately by anaerobic digestion [1].

Microbial Anaerobic Digestion (Bio-Digesters) as an Approach to the Decontamination of Animal Wastes in Pollution Control and the Generation of Renewable Energy © Manyi-Loh CE, Mamphweli SN, Meyer EL, Okoh AI, Makaka G, and Simon M. International Journal of Environmental Research and Public Health *10*,9 (2013). doi:10.3390/ijerph10094390. *Licensed under a Creative Commons Attribution 3.0 Unported License, http://creativecommons.org/licenses/by/3.0/.*

So far, the conventional sources of energy that have provided power for developing and maintaining the technologically advanced modern world are the fossil fuels including coal, oil and natural gas. Yet, fossil resources are finite and their continued recovery and use appreciably impact our environment and affect the global climate due to the emission of greenhouse gases. Moreover, shortening of oil and gas are becoming imminent and to prepare for a transition to more sustainable sources of energy, viable alternatives for conservation, supplementation and replacement must be explored [2]. In this regard, biomass materials have been viewed as a way to expand energy supply, help mitigate growing dependence on fossil fuels and alleviate environmental and health hazards emanating as side effects from the use of fossil resources in many developing and developed countries [3].

Anaerobic digestion of biomass wastes could have a huge impact on renewable energy requirements. It is best suited to convert organic wastes from agriculture, livestock, industries, municipalities and other human activities into energy and fertilizer. It has become popular in developing countries such as China, India and Nepal; however, in South Africa, biogas digesters are principally constructed and installed in the Western and Kwa-Zulu Natal provinces of the country [4]. Owing to the important roles demonstrated by rumen microorganisms in anaerobic digestion [3], animal manures have been established as suitable sources of biogas production in Africa although, they are co-digested with energy crops in Denmark and Germany [5,6]. Co-digestion refers to the simultaneous anaerobic digestion of multiple organic wastes in one digester. This principle enhances methane yield due to positive synergisms established in the digestion medium, bacterial diversities in different wastes and the supply of missing nutrients by the co-substrates [7].

Furthermore, these wastes obtained from different animals vary in chemical composition and physical forms as a result of principal differences in the digestive physiology of the various species, the composition and form of diet, the stage of growth of the animal and lastly the management system of waste collection and storage [8]. Moreover, Sakar et al. [9] and St-Pierre and Wright [10] stated that a large proportion of the agricultural sector in both developing and developed countries is involved with poultry and livestock farming resulting in huge quantities of animal manure and other wastes from livestock operations which merit public, environmental and social concerns. Consequently numerous digesters are designed and installed on farms for the proper management of these wastes [11].

Overall, anaerobic digestion reduces biomass wastes and mitigates a wide spectrum of environmental undesirables, it improves sanitation, helps in air and water pollution control and reduces greenhouse gas emissions. Also, it provides a high-quality nutrient-rich fertilizer and yield energy in the form of biogas. The uses of biogas vary greatly from developing to developed countries. In Africa, biogas generated can be used as fuel for cooking, lighting and heating; it reduces the demand for wood and charcoal for cooking therefore helps preserve forested areas and natural vegetation, and can also help alleviate a very serious health problem due to poor indoor air quality associated with wood and charcoal used for cooking [12,13].

In Western countries (e.g., Germany & America), biogas is converted to electricity and heat for on-farm purposes by combined heat and power units after removing water and sulphur from its mixture [14]. Alternatively, it is upgraded to bio-methane constituting 95–99% methane wherein it opens up more utilization opportunities. Bio-methane is fed into the gas grid and used as power, transportation fuel and for heating [15].

Against this background, this paper appraises insights on environmental and public health implications arising from improper disposal of animal wastes and a comprehensive description of anaerobic digestion of these animal wastes as a means of resolving the ills; with emphasis on types of bio-digesters, microbial communities engaged in the process and techniques for their identification as well as factors affecting the digestion process.

3.2 ENVIRONMENTAL AND PUBLIC HEALTH IMPLICATIONS OF ANIMAL MANURE

Wastes from agricultural animals (poultry and livestock) often contain high concentrations of human pathogens, spilled feed, bedding material, fur, process-generated wastewater, undigested feed residues, feces and urine therefore must be effectively managed to minimize environmental and public health risks. However, the type and pathogenic microbial load depend on the type of the waste and its composition [16]. Figure 1 shows an overview of different contaminants in animal wastes and plausible implications. The following contaminants including pathogens (bacteria, viruses and protozoa), nutrients (phosphorus, nitrogen and sulphur), heavy metals (zinc and copper), veterinary pharmaceuticals (antibiotics) and naturally excreted hormones are present in animal manure [17].

3.2.1 SOURCES OF CONTAMINANTS IN ANIMAL MANURE

The intestinal tract of human and animals have been found to be the major sources of *Salmonella* and *Escherichia coli* in nature [18], which could be shed in feces. These pathogens may persist for days to weeks to months depending on the type of pathogen, the medium and the environmental conditions. Approximately 1% to 3% of all domestic animals are infected with *Salmonella*e [12,19]. Furthermore, other non bacterial pathogens that may be present with fecal material include protozoa (*Cyptosporidium* and *Giardia*) and viruses (Swine Hepatitis E- virus). The management and disposal of animal wastes harboring such pathogens can increase the risk of infections and diseases that threatens human health if these wastes are not properly treated and contained [20].

Antibiotics are routinely used in animal farming to prevent the spread of diseases or treat infected animals and simply added to animal feed to promote/increase growth. Further compounding the problem is the fact that the misuse or overuse of antibiotics could speed up the development of resistance or increase resistance of the microbial population present [17] due to the fact that resistance genes may be transferred between the microbial communities present. The use of any one antibiotic can yield resistance to multiple antibiotics. However, it is devastating that most antibiotics are designed to be quickly excreted from the treated organisms. Thus, they are commonly found in animal wastes.

Consequently, microorganisms in animal manures are thought to affect human health via multiple pathways either through direct or indirect contact with food, water, air or anywhere manure goes [21]. Further transmission of pathogens off-farm from farm workers to family members is also possible. It has been noted that most human-acquired infections result from these resistant strains [22]. Antimicrobial resistance of microorganisms is a local, national and global challenge; therefore the quest to identify the sources of these antibiotic resistant microbes and to seek for means to stop the spread of their resistance genes or not to select for further resistant strains is imperative.

Furthermore, animal manure may cause environmental pollution of water bodies as it is described to be rich in nutrients. Seppälä et al. [23] noted that these wastes harbor both micro and macronutrients including zinc and copper. These metals (Zn & Cu) are micronutrients found in animal manure originating from feed, supplements, antibiotics and water consumed by the animals [9].

3.2.2 ADVERSE EFFECTS OF ANIMAL WASTES ON THE ENVIRONMENT AND HUMANS

Taking into consideration the concentration of contaminants in animal wastes, it does have the potential to pollute land, water and air if containment and treatment do not adequately manage it. Haulage of these contaminants in animal wastes is dependent on the chemical characteristics, soil properties, climatic conditions and crop management practices. It is most probable that rain may wash/flush these wastes into streams, rivers or may cause waste to seep through the soil into underground springs and wells that humans use for sanitation and domestic purposes [24].

From the environmental point of view, excessive nutrients (especially phosphorus and nitrogen) in conjunction with elevated levels of biological oxygen demand (BOD) and chemical oxygen demand (COD) in these water bodies can contribute to algal blooms and cyanobacterial growth thus presenting serious socioeconomic hazards [25]. As a long term effect, they may cause shifts in phytoplankton community structure from desirable species to noxious species by means of holding back the growth of desirable aquatic species. Antibiotics in the soil may affect the natural ecosystem functions such as soil microbial activity and bacterial denitrification [17].

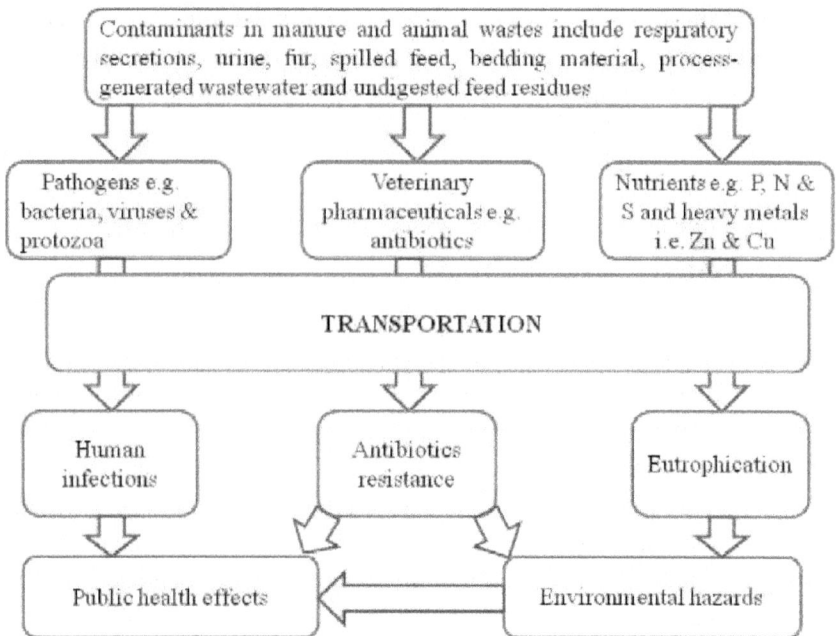

FIGURE 1: Environmental and public health implications of animal manure.

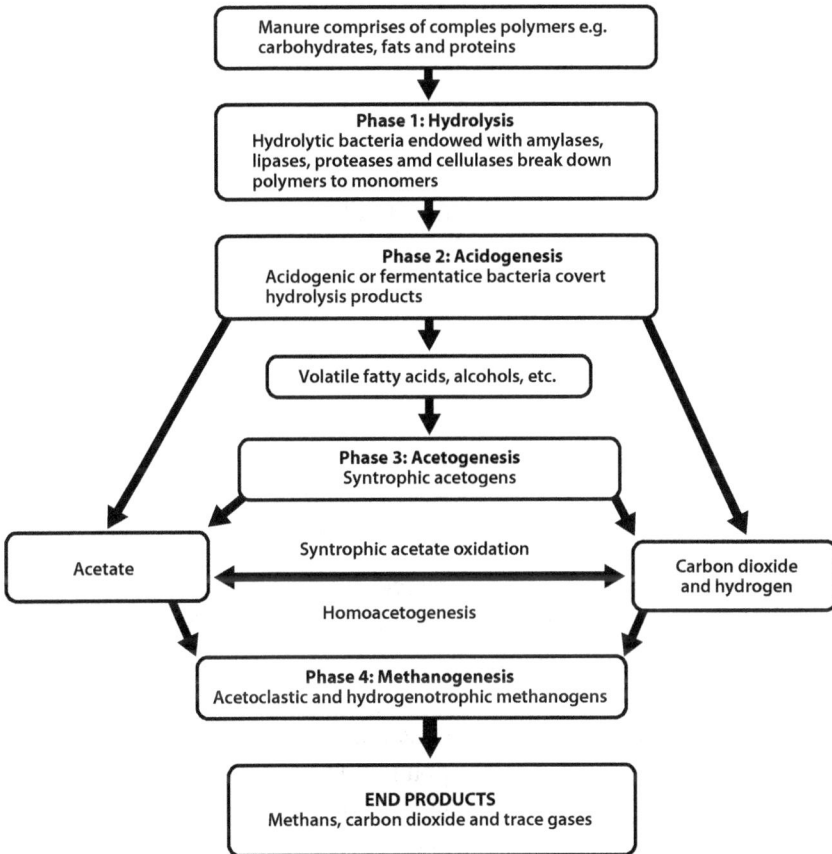

FIGURE 2: A schematic presentation of anaerobic digestion process.

Of profound public health effect is the reality that waterborne pathogens can be accidentally ingested during recreational activities or drinking water that is contaminated with animal feces [26]. This might result
in acute gastrointestinal upset, e.g., nausea, diarrhea and vomiting. Also,
contact with affected surface water during recreational activities can cause
skin, ear or eye infection. Cyanobacteria (blue green algae) in surface water can produce neuro-toxins and hepatotoxins which are known to cause
acute and chronic health complications [27]. However, the disease conditions in susceptible individuals including the very young, the elderly, preg-
nant women, and immunocompromised may be more severe, chronic or
at times fatal [26]. In addition, antibiotic resistance interferes with antimicrobial chemotherapy causing treatment failure of some infections thereby
presenting life threatening situations. Consequently, anaerobic digestion
of animal manure in bio-digesters can substantially help to address the
above-mentioned troubles arising from the improper or careless disposal
of these wastes.

3.3 ANAEROBIC DIGESTION OF ANIMAL WASTES
IN BIO-DIGESTERS

Growth and intensification of livestock operations often result to great
quantities of manure that have to be properly managed. Even when stored,
manure generates and releases methane (a greenhouse gas) into the atmosphere [10]. Moreover, anaerobic degradation has usually taken place
in the lower digestive tract of animals and then continues in the manure
piles resulting in malodorous compounds. These malodorous compounds
originate from the incomplete breakdown of organic matter in manure by
anaerobic microbes under uncontrolled environmental conditions [28].

Farm-based anaerobic digester presents as an alternative to the proper
management of these wastes [29]. There are thousands of on-farm digester
plants worldwide including Blue Spruce Farm, Green Mountain Diary,
Chaput Family Farm, Cantabria Diary Plant, Buttermilk Hall Farm, Bul-
cote Farm, Minnesota mid-sized Diary Farm and a host of others [10,30].
However, Lutge and Standish [31] noted that very few on-farm anaerobic
digesters are available in South Africa. Nevertheless, the type of digester

used is being influenced by the characteristics of manure collected which is dependent on the animal's diet and on-farm practices [32]. All the same, these farms employ anaerobic digestion which is a consistent technology for their waste management/treatment.

Taking into consideration the aforementioned differences in livestock practices existing between the farms, the essential step of liquid-solid separation of manure mixture may be performed before or after anaerobic treatment [33]. In addition, the manure collected may be mixed with milk house waste for anaerobic digestion [34]. Generally, manure is collected with or without milk house waste and slurry prepared by adding water to it. The slurry is pumped to the separator for screening, separating the mixture into liquid and solid fractions. Subsequently, the screened liquid fraction is fed into the digester whilst the solid fraction could be dewatered and redistributed to areas lacking nutrients, used as bedding and or composted to serve as an additional source of more carbon and nitrogen [9]. In addition, the digested liquid fraction can be processed to obtain concentrated fertilizers or post-treated to obtain clean water for recycling and irrigation purposes [30,32].

Overall, during anaerobic digestion complex polymers in animal wastes are catabolized through a series of steps by complex consortia of microorganisms in the digester to ultimately yield methane and carbon dioxide [35]. Basically, this process can be divided into four phases: hydrolysis, acidogenesis, acetogenesis and methanogenesis in which hydrolytic, fermentative bacteria, acetogens and methanogens play distinct roles, respectively, as shown in Figure 2.

During hydrolysis, complex polymers like carbohydrates, proteins and fats are being degraded into sugars, amino acids and long chain fatty acids respectively. This breaking down process occurs primarily through the activity of extracellular enzymes (lipases, proteases, cellulases & amylases) secreted by hydrolytic bacteria attached to a polymeric substrate [36].

Later, fermentative or acidogenic bacteria transform hydrolysis products into acetic acid and intermediate compounds, such as ethanol, lactic acid, short chain fatty acids (C3–C6), hydrogen and carbon dioxide [35]. Acetate, carbon dioxide, formate, methylamines, methyl sulphide, acetone and methanol produced in this phase can be directly utilized for methanogenesis. Consequently, the other intermediary products from acidogenesis are converted to acetate, formate or CO_2 & H_2 by syntrophic acetogens in a bid to maximize methane production.

As a final point, methane is produced during methanogenesis by methanogens in two ways: either through cleavage of acetic acid molecules to produce methane and carbon dioxide or reduction of carbon dioxide with hydrogen by acetotrophic and hydrogenotrophic methanogens, respectively [37]. The biogas generated constitutes mainly methane (50–75%), CO_2 (25–45%) and traces of other gases like CO, H_2S, NH_3, O_2, water vapor (Table 1).

TABLE 1: Various constituents of biogas generated from the anaerobic digestion process; its average composition adopted from de Graaf and Fendler, [38].

Component	Symbol	Percentage
Methane	CH_4	50–75
Carbon dioxide	CO_2	25–45
Hydrogen	H_2	1–2
Ammonia	NH_3	<1
Water vapor	H_2O	2–7
Oxygen	O_2	<2
Hydrogen sulphide	H_2S	<1

3.3.1. MICROBIAL COMMUNITIES INVOLVED IN ANAEROBIC DIGESTION OF ANIMAL MANURE AND METHODS OF THEIR IDENTIFICATION

The specific microbes and their metabolic activities during anaerobic digestion depend on the chemical composition of the feedstock/waste, environmental factors and digester operating conditions [39,40]. Four sets of microorganisms are involved and these groups of microorganisms are tightly attached metabolically whereby the early stages of digestion yield reduced intermediate products that are utilized by acetogens and methanogens [37]. However, the interrelationship between the acetogens and methanogens is highly complex. These microorganisms are classified as anaerobes therefore oxygen poses a threat via the disruption of metabolic pathways causing oxidation of cellular factors that normally occur in reduced form. Contrarily, in recent times, it has been documented that several methanogens adapt to oxygen due to the presence of genes that synthesize enzymes (e.g., catalase and superoxide dismutase) within their genomes, which serve a role in defense against oxygen toxicity [41]. Several authors have reported the high tolerance of methanogens including *Methanobacterium thermoautotrophicum, Methanobrevibacter arboriphilus* and *Methanosarcina barkerii* to oxygen and dessication [42,43].

More elaborately, Anderson et al. [44] also noted that after dessication process, *M. barkeri* had innate capability to survive extended periods of exposure to air and lethal temperatures owing to the synthesis of thick outer cell layers composed of extracellular polysaccharide (EPS); added to the accumulation of cyclic 2,3-diphosphoglycerate (a novel metabolite which may be used to stabilize proteins at elevated temperatures). In addition, the membrane lipids of archael species have glycerol molecules bound by ether linkages to branched isoprene hydrocarbon molecules causing the organisms to adjust to such extreme temperatures [45]. On the whole, microbial community within a digester system can be grouped into acidogens, syntrophic acetogens and methanogens [46].

3.3.1.1 ACIDOGENS

It has been documented that the bacterial species active in the polymer hydrolysis phase are also active during the acidogenic phase. Hence, the hydrolytic and acidogenic bacteria are sometimes referred to as fermentative bacteria. They can be either facultative anaerobic bacteria (i.e., can survive under both aerobic and anaerobic conditions) or strict anaerobes. The family *Enterobacteriaceae* or enteric bacteria (a group of bacteria that inhabit the intestine of humans and other animals) are active fermenters and are among the organisms responsible for the first step in the bioconversion of carbohydrates to CH_4 [18].

In addition, Blumer-Schuette et al. [47] and Wirth et al. [48] documented the relevance to biomass deconstruction of the following microorganisms: *Caldicellulosiruptor saccharolyticus, Thermotoga maritima, Clostridium thermocellum, Anaerocellum thermophilum, Escherichia coli, Clostridium kluyveri, Bacillus cereus, Ruminococcus albus* etc. Other known groups of anaerobic cellulose-degrading bacteria are found in the following genera; *Aminobacterium, Psychrobacter, Anaerococcus, Bacteroides, Acetivibrio, Butyrivibrio, Halocella, Spirochaeta, Caldicellulosiruptor* and *Cellulomonas* (a facultative anaerobe of the phylum *Actinobacteria*) [49,50].

3.3.1.2 SYNTROPHIC ACETOGENS

Syntrophic acetogens, e.g., *Syntrophobacter wolinii, Syntrophomonas wolfei* and *Smithella* sp. are responsible for the syntrophic metabolism of alcohols, short chain fatty acids (C3–C6), some amino acids and aromatic compounds to yield methanogenesis substrates [51]. The conversion of the above-mentioned substrates to yield methanogenesis products is thermodynamically unfavorable but it becomes favorable with the presence of a syntrophic partner (hydrogenotrohphs) [51,52].

However, accumulation of volatile fatty acids results in decrease in pH, increase acidification, destroy methanogens activity and leads to failure of digester ultimately [7]. While the syntrophic acetogens are converting intermediary metabolites to acetate and other methanogenesis substrates, homoacetogens also produce acetate from the reduction of carbon dioxide with hydrogen via the acetyl Co-A pathway [53].

Overall, methane is produced during methanogenesis by acetotrophic, hydrogenotrophic and methylotrophic pathways. In regards to acetate degradation to yield methane via the acetoclastic pathway, specific methanogens from the order *Methanosarcinales* are responsible. Contrarily, a group of acetate-oxidizing bacteria occur in syntrophic relationships with hydrogenotrophic methanogens wherein they oxidize acetate to form methane in association with the latter microorganisms.

These bacteria are equally referred to as syntrophic acetogens and include both mesophilic and thermophilic bacteria (Table 2). Syntrophic acetate-oxidizing bacteria are involved in the reversed reductive acetogenesis [54] and can be identified by the combinatorial use of flux measurement and transcriptional profiling of formyltetrahydrofolate synthetase (FTHFS) gene, an ecological biomarker engaged in reductive acetogenesis [52].

TABLE 2: Syntrophic acetate-oxidizing bacteria in association with hydrogenotrophic methanogens.

Acetate-oxidizing bacteria	Microbial description	Hydrogenotrophic methanogens	References
AOR	Anaerobic, rod-shaped, gram positive, non-spore forming and thermophilic (60 °C)	*Methanobacterium* sp. strain THF	Lee and Zinder [55]
Clostridium ultunense	Anaerobic, spore-forming, rod-shaped, gram negative and mesophilic (37 °C)	*Methanoculleus* sp. strain MAB1	Schnürer et al. [56]
Thermacetogenium phaeum	Anaerobic, rod-shaped, gram negative but with gram positive cell wall structure and thermophilic (between 55 and 58 °C)	*Methanothermobacter thermoautotrophicus* TM	Hattori et al. [57]
Thermotoga lettingae	Anaerobic, rod-shaped, non-spore forming, mobile, gram negative and thermophilic (65 °C)	*Methanothermobacter thermoautotrophicus* or *Thermodesulfovibrio yellowstonii*	Balk et al. [58]
Syntrophaceticus schinkii	Anaerobic, spore-forming, variable cell shape, gram variable and mesophilic (between 25 and 40 °C)	*Methanoculleus* sp. strain MAB1	Westerholm et al. [59]

3.3.1.3 METHANOGENS (ARCHAEA)

Methanogens are found in a wide range of anaerobic habitats including freshwater and marine habitat, sewage digesters, the digestive tracts of herbivores, mammals and wood and humus feeding insects etc. [60,61,62,63]. They belong to the domain *Archaea* and they occupy a key position in the anaerobic digestion process because it is in this last step of the process where the valuable methane is produced [45]. During an unstable anaerobic digestion process in a poorly performing anaerobic digester, the methanogenic populations are especially sensitive to acidity (pH), concentrations of volatile fatty acids, and free ammonia and ammonium ions in the digesting substrate [64].

In addition, methanogens are classified into six orders i.e., *Methanobacteriales, Methanococcales, Methanomicrobiales, Methanosarcinales, Methanocellales* and *Methanopyrales* [65]. Members of the order *Metha-*

nosarcinales utilize acetate which has for a long time been known as the major precursor for more than 70% of methane produced in most engineered anaerobic digester [66]. In other words, previous knowledge assumed that in totality, two-thirds of methane are obtained from the acetoclastic methanogenesis pathway and one-third from hydrogenotrophic methanogenesis pathway according to anaerobic digestion model 1 (ADM1) [66]. This is misleading since reports were based on anaerobic digestion of waste water and sewage sludge with very low organic content.

Clearly, biomethanation involves a complex community of specialized microorganisms that depend on each other either for substrate supply or metabolism of end products in order to favor their metabolic activity. Moreover, microbial species require specific combination of physical and chemical conditions viz temperature, pH, salinity besides substrate availability to thrive. Therefore, microbial species obtained from different environment associated with different physical and chemical factors tend to vary even though they could perform anaerobic digestion through the same stages of hydrolysis, acidogenesis, acetogenesis and methanogenesis [10]. However, this might influence the biogas yield obtained under different physical, chemical and substrate conditions.

Nowadays, results obtained from several studies conducted by different authors have contradicted the assumptions of ADM1. Krakat et al. [67,68] demonstrated that a higher temperature of 60 °C and a drastic reduction in hydraulic retention time resulted in dominance of hydrogenotrophs among the microbial communities with a corresponding increase in methane production in thermophilic and mesophilic biogas fermentors respectively, digesting energy crops. Similarly, Klocke et al. [69,70] reported varying substrate utilization during methanogenesis within the biogas plant as depicted from the dominance fraction of hydrogenotrophic methanogens relative to acetoclastic counterparts.

Furthermore, members of *Methanosarcinales* are described as acetoclastic and comprise of two families *Methanosarcinaceae* and *Methanosaetaceae*. However, these two families of acetoclastic methanogens differ in their physiology, biokinetics and growth environment with respect to the acetate concentration [2]. Conclusively, the interactions of the different groups of anaerobic microorganisms are incredibly complicated, and the effective performance of the biological process strongly depends on the balance of these relationships [71].

3.3.2 TECHNIQUES FOR IDENTIFYING MICROORGANISMS INVOLVED IN THE ANAEROBIC DIGESTION PROCESS

St-Pierre and Wright [10] mentioned that the microbial communities within an anaerobic digester treating animal manure are not fully characterized. However, due to the diversity of microorganisms in the system, a variety of methodological approaches are required for a detailed analysis of community structure in a bid to unravel the complex antagonistic and synergistic effects between microbial communities in order to eventually improve process stability and efficiency of biogas formation [72,73]. This can be achieved by the combined use of traditional culture-based, microscopic and molecular techniques.

Conversely, culture-based techniques such as plate counts, membrane filtration and most probable number (MPN) have an inherent limitation because only the viable population will grow to produce colonies under specific growth conditions whereas others that are important in the original sample do not proliferate [74]. However, these traditional culturing methods employed with environmental samples also underestimate the total number of microorganisms due to the selective nature of the media used, the lack of detection of active but non cultivable (ABNC) microbes and failure to count microbes that are present as aggregates or associated with particles. Likewise, it is impossible to obtain pure cultures of most microorganisms in natural environment due to the complex syntrophic and symbiotic relationships that are abundant in nature [75].

Contrarily, direct microscopic methods e.g., DAPI epifluorescence microscopy allows the direct observation and total enumeration of viable and non-viable microorganisms in the feedstock [72]. Specifically, the identification and enumeration of methanogenic microbes can be achieved by epi-fluorescence microscopy, a technique based on their unique fluorescent pigment, factor F_{420} [76]. The coenzyme F_{420} shows autofluorescence (blue-green color) of methanogenic cells when excited by UV light. Hence, this autofluorescence serves as a diagnostic tool used to count autofluorescent methanogens [77]. Results from this technique can be combined with those from molecular methods for a better overview of microbial population and structure in an anaerobic digester [78].

Molecular techniques targeting particularly 16S rRNA genes (the only RNA component of 30S ribosomal subunit), can also be employed as conventional methods for identification of microbial community in a digester. These methods include cloning, fluorescent in situ hybridization (FISH), denaturing gradient gel electrophoresis (DGGE), single strand conformation polymorphism (SSCP), restriction fragment length polymorphism (RFLP), quantitative real-time PCR (qPCR) and DNA sequencing (Sanger and next generation sequencing methods). Each of these methods has its advantages and shortcomings that have been deliberated and presented elsewhere [74,79].

Most previous studies on anaerobic digestion of biomass materials were based on the construction of 16S rRNA clone libraries and subsequent sequencing (Sanger method) of individual clones [69]. The resulting sequences corresponded to different taxa that were employed in the phylogenetic classification thereby revealing the structure of the underlying community. In addition to knowledge on phylogenetic diversity, the microbial population present in an anaerobic digester can be quantified using FISH and qRT-PCR as well as uncultured microbes will be uncovered by the use of molecular methods based on 16S rRNA [80]. However, the analysis of 16S rRNA gene does not cover the whole complexity of the environment due to sequencing of limited number of clones as well as low cloning efficiencies [81].

In addition to the methods relying on 16S rRNA for the detection of microorganisms in an anaerobic digester, other methods are available incorporating other remarkable genes which could serve as a diagnostic tool. The hydrazine oxidoreductase genes (*hzo* gene) can be employed in phylogenetic diversity and functional analysis of anaerobic ammonium-oxidizing bacterium (anammox) in a community or an environment. The hzo genes are specific for the identification of anammox bacteria and can be used as a functional marker to give activity-based information regarding the anammox bacteria in a community [82]. Ozgun et al. [83] identified and quantified anammox bacteria from waste water treatment plant by adopting a FISH-based approach with the PCR primers designed for the amplification of *hzo* specific genes.

Furthermore, hydrazine synthase (hzs A) also represents a unique phylogenetic marker for anammox bacteria hence; it can be used for identification purpose. Moreover, Harhangi et al. [84] determined the presence and biodiversity of anammox bacteria from samples of wastewater treatment systems, fresh water and marine environments as well as anammox enriched cultures by the development of PCR primer set targeting a subunit of hydrazine synthase. The tested primers successfully retrieved hzs A gene sequences covering all known anammox genera thus indicating that the use of 16S rRNA gene does not directly relate to the physiology of the targeted microbes and also the primer set currently available does not give tangible information pertaining to the diversity of the anammox species.

Also, methanogens can be distinguished from other microbes by virtue of their cell wall components, unique membrane lipids, 16S rRNA gene sequences as well as key enzymes that are involved in methanogenesis. The key enzymes encompass specific coenzyme and cofactors such as F_{420}, methanopterin and coenzyme M that are engage in methanogenesis [85]. These enzymes are strongly conserved and found only among methanogens. Consequently, methanogens can be specifically identified by targeting genes that code these peculiar enzymes.

Of great significance is the methyl Coenzyme M reductase (an enzyme complex) that catalyzes the reduction of the methyl group bound to coenzyme M and is encoded by the gene *mcr A*. It has been documented that methanogenic biodiversity displayed upon utilization of *mcr A* is similar to that revealed by 16S rRNA therefore validating the application of the former gene for identification [86,87]. However, a prominent drawback linked to the incorporation of *mcr A* gene is that its available sequences are very limited in the database therefore primers designed based on these sequences may lead to inefficient amplification and an adequate representation of the methanogenic community will not be delineated [85].

Interestingly, next generation sequencing methods (NGS) are emerging as a robust technology that creates abundant data and has caused a fundamental shift in molecular biology. These new methods involve whole genome sequencing; function-driven metagenomics and high throughput sequencing [85] that have the potential to demonstrate new insight into the entire genome of the microbial environment resulting in the rapid characterization of targeted sequences at less cost compared to the first-generation sequencing method, "traditional Sanger method" [88].

Sequencing of the whole genome is often aimed at obtaining information about the complete set of genes in any particular genome [89]. Genome sequencing opens up unexpected opportunities with information that cannot be obtained with conventional microbiological methods. For instance, Leahy and colleagues [90] unravelled the presence of vaccine and chemogenomic targets in their attempt to assess the sequencing of the M1 genome (first ever identified genome of rumen archael species) of *M. ruminatium*. Attwood et al. [61] equally noted the presence of a phage sequence residing within the genome of *M. ruminatium* and postulated that enzymes from the phage could serve as potential methanogen control agents. However, knowledge of the complete genetic makeup of an organism is not sufficient as information pertaining to their function is missing thus indicating that the biochemical function of each gene product is crucial.

On the other hand, metagenomics or community genomics or environmental genomics is the analysis of genomic DNA of the whole microbial communities directly in their natural environments by extracting or isolating the total DNA from an environmental sample [91]. Specifically, function-driven metagenomics is a powerful tool that has the potential to identify entire new classes of genes of new or known functions thus providing information on the metabolic activities of all members of a microbial community, including even those that were uncultivable and undefined previously [91].

In addition, Gilbert and Dupont [92] further described that metagenomics embodies two aspects; the environmental single-gene surveys in which single targets are amplified and the PCR- amplified products are sequenced and secondly, random shotgun studies of all environmental genes where the total DNA is isolated and sequenced; giving a profile of all the genes occurring in the community. However, the extent of community coverage depends on the depth of sequencing.

With the advent of high throughput sequencing technologies such as Roche 454, Illumina/Solexa, Applied Biosystems/SOLiD and Helicos BioSciences, more sequence data are obtained than has ever been possible with the traditional Sanger sequencing method [10]. These improved sequencing methodologies do not require cloning of the DNA before sequencing thereby; they bypass one of the main biases in environmental sampling. Consequently, bioinformatics strategies and tools can be used for the analysis of the huge data obtained [85].

In conclusion, conventional molecular tools based on 16S rRNA (viz PCR, qRT-PCR, FISH, DGGE, cloning and sequencing of gene library) are still of relevance since they provide an initial selection of collected samples prior to comprehensive analyses by next generation sequencing technologies [85]. Moreover, from clone libraries of 16S rRNA, usually precise sequences are assigned as a result of larger sequence length relative to the short read lengths obtained from high throughput sequencing technologies [50]. Therefore, it is very promising integrating conventional molecular methods, epi-fluorescence microscopy and high throughput sequencing technologies in describing the microbial structure of a biogas plant. Combining these methods will reveal the entire complexity of the microbial communities in an anaerobic digester as well as the physiology and function/activity of the community. Also, previously unidentified microbes with culture-based techniques will be delineated and their function within the community will be recognized [91].

3.3.3 TYPES OF BIO-DIGESTERS FOR TREATING ANIMAL MANURE

A biogas digester consists of one or more airtight reservoirs (chambers) into which animal manure or a mixture of manure and co-substrate is placed, either in batches or by continuous feed [93]. These biogas generating systems could be categorized on the basis of the number of reactors used into single (one) stage or multi (two) stage and on the mode of feeding into continuous and batch feeding systems [12].

In single stage processes, the three stages of anaerobic process occur in one reactor; however, the growth rate of fermentative bacteria is faster than that of acetogenic and methanogenic bacteria [33]. Consequently, acids accumulate; the pH falls and the growth of methanogenic bacteria is inhibited due to increase organic loading rate and inappropriate other process parameters. Whereas multi-stage processes make use of two or more reactors that separate the acetogenesis and methanogenesis stages in space and allows the establishment of operational conditions that reduce the start time and microbiota specialization in each reactor, thereby allowing the most desirable products at each stage to be obtained [35,69].

In a batch experimental set up, the digester is loaded with the feedstock at the beginning of the reaction and the product discharged at the end of each cycle whereas in continuous feeding, the organic material is continuously charged and discharged [12].

Livestock operational practices differ between individuals and they influence the characteristics of the manure obtained which in turn determines the choice of the digester [28]. Manure can be collected either by scraping with an automated device or flushing with water [32]. Ideally, scraped manure can be digested by a complete mix digester (e.g., continuously stirred tank reactor, CSTR) and a plug flow digester whereas flushed manure warrants the use of covered lagoons and anaerobic fixed film digesters [31,94].

Traditionally, animal manure is often flushed, pretreated by means of mechanical screening, sedimentation or both in a bid to achieve two separate fractions of liquid and sludge; with the liquid portion pushed into covered lagoons for storage and anaerobic treatment [28]. However, the anaerobic digestion process in the lagoon is affected by climatic conditions (temperature) as well as the water table on the site especially since liquid can seep into underground spring and streams [95]. In recent times, with the quest for retaining active microbial population in the digester in order to improve process stability and control; biodigesters are designed with active microbial populations attached to inert supports as biofilms or form aggregates or granules [9,94]. Sakar et al. [9] revealed that up flow anaerobic sludge blanket is most suitable for the anaerobic digestion of poultry wastes.

More elaborately, conventional digesters used for anaerobic treatment of animal manure are CSTR and plug-flow reactors with appreciable holding capacity though with long HRT compared to fixed film digesters [11]. On the other hand, anaerobic fixed film reactors have the potential of retaining microbial mass (as biofilms) on support materials and also reduce the retention time for anaerobic digestion to several hours or a few days [94]. However, Lutge and Standish [31] mentioned that South Africa has the potential of implementing the use of CSTR and covered lagoons for on-site animal manure treatment.

3.3.4 FACTORS INFLUENCING ANAEROBIC DIGESTION OF ANIMAL MANURE

Generally, factors affecting the performance of an anaerobic digester include operational factors (pH, temperature, organic loading rate (OLR)/ hydraulic retention time (HRT), free ammonia concentration), substrate characteristics/biodegradability and biodigester design [96]. However, Wilkie [28] reported that temperature, biodegradability, OLR and HRT have great impact on the anaerobic digestion of animal manure. Notwithstanding, other factors should not be overlooked.

3.3.4.1 TEMPERATURE

Based on temperature, anaerobic microorganisms can be categorized into psychrophiles (<20 °C), mesophiles (25–37 °C) and thermophiles (55–65 °C) [97]. Some methanogenic species exhibit a preference of extreme heat (90–100 °C) thus, are classified as hyperthermophilic methanogens [45]. Examples are *Methanocaldococcus jannaschii* and *Methanococcus vulcanius* [62]. Temperature can be considered as the most important environmental factor influencing the growth of microbes. Albeit, each microorganism has a certain temperature range within which it can grow and multiply. When temperature is increased within a certain range, the chemical and enzymatic reactions increase at a faster rate and growth increases [98].

However, above optimum temperature, key chemical reactions in the different metabolic pathways being catalyzed by enzymes cannot occur because enzymes are irreversibly destroyed since they are protein like in nature. Enzymes are crucial to metabolism because they allow organisms to drive desirable reactions that require energy and will not occur by themselves, by coupling them to spontaneous reactions that release energy. Consequently, the growth rate of microbes will equally stop [99]. However, different microbial species respond differently to abrupt changes in temperature. Moreover, temperature does not only influence the rate of metabolism of the microorganisms but also affect other process parameters such as OLR and ammonia concentration [29,100].

Generally, anaerobic digestion of biomass wastes could occur both at mesophilic (25–37 °C) and thermopilic (55–65 °C) temperature ranges. However, the ratio of free ammonia to total ammonium ion is higher at thermophilic temperature ranges. Consequently, animal wastes (containing nitrogen and ammonia compounds) are digested at mesophilic temperature range (25–37 °C) in a bid to avoid ammonia mediated inhibition of methanogenesis [29]. Moreover, thermophilic treatment requires high energy thereby may reduce the net energy obtained from the overall digestion process [100]. In spite of the abovementioned drawbacks of thermophilic fermentation; it undoubtedly causes significant destruction of pathogens and weed seeds and also causes higher metabolic rate resulting in higher methane yield [100,101].

3.3.4.2 PH AND ALKALINITY

In regards to anaerobic digestion, it is more appropriate to discuss pH alongside alkalinity since the latter can be used to control pH thus buffering the acidity of the system derived from acidogenesis phase [102]. Therefore, the amount of alkalinity present in an anaerobic digester represents the buffering capacity.

The pH range of anaerobic digestion normally occurs near neutral pH range and it is dependent on the OLR (which depends on reactor type) and the buffering capacity of the substrate. Livestock wastes (rich in ammonia and nitrogen compounds) such as cow, swine and poultry manure have high buffering capacity as they produce alkalinity when degraded upon by microorganisms [103]. However, anaerobic digestion of these wastes is often maintained at higher pH values of 7.6 [29]. An increase in OLR with a corresponding decrease in HRT can result to accumulation of volatile fatty acids which causes a drop in pH due to increased acidity of the digesting medium [104]. However, in instances where the pH has to be adjusted, several chemicals such as sodium hydroxide, potassium hydrogen carbonate, sodium carbonate, calcium carbonate, calcium hydroxide etc. can be added for alkalinity supplementation [40].

3.3.4.3 AMMONIA CONCENTRATION

Anaerobic digestion of urea- and protein-rich wastes such as animal wastes is often faced with the challenge of high levels of free ammonia due to their high organic nitrogen concentration which upon biological degradation results in high concentration of total ammonium ion plus free ammonia [105]. The quantity of ammonia produced during the digestion process is attributed to substrate concentration of nitrogen, reactor loading, C/N ratio, buffering capacity and temperature. In aqueous solution, inorganic ammonia nitrogen exists in two principal forms; ammonium ion (NH_4^+) and unionized ammonia or free ammonia (NH_3) in a pH dependent equilibrium state. Ammonia toxicity is influenced by the operating pH and temperature [29].

An increase in pH will cause increase ammonia toxicity of the system since a greater part of the total ammonia nitrogen will be free ammonia, the form that has been recognized as a toxic agent [106]. On the other hand, reduction in pH to a level within the optimum pH necessary for growth of the microorganisms will help to counteract free ammonia concentration. However, process instability provoked by ammonia toxicity often results in increased level of volatile fatty acids with a corresponding decrease in methane yield [64].

In addition, Strik et al. [107] noted that high ammonia concentration led to poor biogas quality requiring treatment, decreased COD removal efficiency, decreased biogas generation and malodor, besides process inhibition. High free ammonia content has usually been associated with unstable process performance and increased risk of process failure as a result of its inhibitory effect on methanogens (specifically acetate-utilizing methanogens). Therefore, in the presence of elevated levels of ammonia in a fermentor, a shift occurs in the biomethanation process from aceto-clastic methanogenesis (performed by acetate-utilizing methanogens) to syntrophic acetate oxidation conducted by syntrophic acetogens in collaboration with hydrogenotrophs [64]. Moreover, El-Mashad et al. [100]

revealed that ammonia toxicity does not only affect the acetoclastic methanogens but also hydrolysis and acidification processes.

Furthermore, chemical equilibriums especially of free ammonia concentration at a fixed total ammonium concentration can be affected by the operating temperature (i.e., mesophilic and thermophilic temperature ranges) of the digester system. Even though the temperature is pivotal in the thermodynamics and kinetics of microbial reactions in methanogenesis, challenges are encountered during treatment at thermophilic temperatures (i.e., 55–65 °C) of ammonium-, urea- and protein-rich biomass materials owing to a high level of free ammonia [108]. Garcia et al. [29] noted that at higher temperatures the ratio of free ammonia to total ammonium was much higher; consequently affecting methane generation due to free ammonia inhibition. Nevertheless, an increase in temperature within the mesophilic range relieved the digester system of ammonia toxicity. Therefore, anaerobic digestion of animal manure at mesophilic temperature offers better process stability and performance of the digester system than at thermophilic temperatures [101].

It has been observed that co-digesting animal wastes with carbon-rich co-substrates will help to prevent both volatile fatty acid and ammonia mediated inhibition [103]. However, the inhibitory ammonia threshold concentration is not standardized because of the conflicting results obtained from different studies conducted under different environmental conditions with different substrates and inocula in conjunction with the complex nature of the anaerobic digestion process and acclimation periods [106]. The way microbes tend to adapt to increased ammonia level is dependent on the rate of ammonia formation which is influenced by the OLR and HRT.

3.3.4.4 HYDRAULIC RETENTION TIME
AND ORGANIC LOADING RATE

HRT is the average period of time that the substrate resides in the anaerobic digester and OLR describes the amount of organic matter expressed in g COD/L or g TS/L or g VS/L added to the digester per reactor volume and unit time. HRT is inversely proportional to OLR and both are very useful parameters that contribute knowledge on design and performance of the reactor [109].

Biological decomposition of animal manure is affected greatly by its retention time in the reactor [110]. Retention time is determined by solid content of manure, temperature as well as the type of reactor used for treatment [11]. More elaborately, CSTR and plug flow reactors for animal manure treatment require retention time of 20–30 days whereas fixed film reactors usually have a shorter retention time of several hours to a few days [101]. However, covered lagoons require a longer retention time of 60 days [28]. In addition, HRT also affects the quality of effluent in terms of microbial load, nutrient content as well as the methane yield. Umaña et al. [94] investigated the influence of HRT on anaerobic fixed bed reactor and noted that the quality of effluent and methane yield increased due to increase in HRT.

On the other hand, OLR is dependent on temperature and HRT. An abrupt increase in OLR causes system failure attributed to decreased COD removal efficiency, methane production rate and pH [111]. More elaborately, a higher OLR beyond the optimum capacity elevates the rate of production of intermediary products (fatty acids) by hydrolytic and acidogenic bacteria. Subsequently, these fatty acids would accumulate due to the slow rate of their consumption by methanogens thus; pH will drop thereby inhibiting methanogenic activity [112]. Furthermore, Rincòn et al. [113] documented that a higher OLR influenced the bacterial community within the digester system; with the genus *Clostridium* being predominant at low OLR and the classes and phyla; *Gammaproteobacteria, Deferribacteres, Actinobacteria* and *Bacteroidetes* respectively, predominated at high OLR.

3.3.4.5 SUBSTRATE CHARACTERISTICS AND HEAVY METALS

The constituents of manure directly determine the biogas yield and the level of biochemical reactions that would take place within the digester system [28]. The composition of the manure will depend on the livestock operations which includes the diet and the handling/storage procedure of the wastes [8]. Evidently, for the proper functioning and continuous repro-duction of microbes implicated in the anaerobic digestion process, there is a need for available sources of energy; carbon for the synthesis of new cellular materials, inorganic elements such as nitrogen, phosphorus, potassium, sulfur, calcium and magnesium as well as organic nutrients [114]. As a consequence, the physical and chemical characteristics including the moisture content, total solids content, volatile solids content, phosphorus, nitrogen and carbon content of the feedstock must be evaluated before commencement of the digestion process [40].

Volatile solids of manure are a very critical parameter as it consists of the biodegradable portion which includes carbohydrates, fats and proteins and the refractory portion which cannot be anaerobically digested described as lignocellulosic [28]. The term biodegradability of manure is indicated by biogas or methane yield and percentage of solids (total or volatile solids) that are destroyed in the anaerobic digestion process [115].

Microorganisms require a trace amount of some metals (nickel, cobalt, copper, iron, zinc, molybdenum etc.) for optimum growth and performance. Matseh [116] noted that these trace elements are usually known as stimulatory micro-nutrients and do occur in coenzymes and cofactors. The stimulatory effects potentiated by these metals on biogas process performance are linked to increased methane production, substrate utilization and reactor stability. However, there are wide ranges in the quantity of these metals that are needed in order to become stimulatory; this has been ascribed to differences in pH, OLR, HRT, substrate characteristics and the complex chemical and biological processes monitoring trace metal bio-availability [117].

Moreover, the stimulatory effect varies between the different types of trace metals. This is affirmed by the work of Pobeheim et al. [118] which recorded an increase in methane yield upon addition of a well-defined trace element solution composed of Co, Ni and Mo. Whereas a higher decrease in methane generation and process stability was noted due to the elimination of Ni from the solution. Results further revealed that 0.4–2 μM concentration of Co caused a 10% increase in methane production but the addition of Mo exhibited no profound effect on methane production.

Interestingly, animal manure has been reported to contain a good level of both macro and micronutrients (trace elements) [23]. However, process failure caused by trace element deficiency has been demonstrated during anaerobic digestion of single substrates such as maize silage. Some mono substrates (e.g., maize silage, potato etc.) or even food wastes cannot provide both the micro and macro nutrients essential for the growth of anaerobic microbes that are present in the anaerobic digestion process [119]. Therefore, they have to be supplemented with these nutrients before commencement of the digestion process. Better still, they can be co-digested with animal manure such that animal manure provides good buffering capacity and required nutrients whilst the energy crop provides increases in the energy yield of the process [23,119].

A deficiency of these metals can cause shifts in microbial community structure; Gustavsson et al. [117] documented a shift in microbial community structure from *Methanosarcinales* dominance during a stable process performance with Co and Ni supplementation to *Methanomicrobiales* dominance at both Co and Ni deficiency. However, too high a concentration of these heavy metals would lead to toxicity of the system thereby hampering the biological process via the enzymes involved by interfering with their function and structure. Apparently, they may substitute for naturally occurring metals in an enzyme prosthetic group or by binding to the SH group of the enzyme [106].

3.3.4.6 MIXING

Of great value in the anaerobic digestion of animal manure is the extent of contact between the incoming animal manure and a viable bacterial population; this is a function of mixing in the reactor [110]. The benefits of mixing digester contents during anaerobic process have been documented by several authors and include: it prevents scum formation inside the digester, ensures uniform distribution of microorganisms and substrate throughout the mixture and intensifies contact between them, prevents stratification within the digester therefore enables uniform distribution of heat throughout the mixture and lastly helps to release gas from the mixture [30,120].

According to Ghanimeh et al. [121], stirring can result in reduction of particle sizes of the substrate as anaerobic digestion progresses. However, what is unclear about the aspect of mixing is the intensity and the duration of mixing considering the fact that different modes (mechanical mixers and recirculation pumps) could be used [110]. In the characterization of manure, total and volatile solids are very paramount because there is a certain limit above which the manure will no longer be a slurry hence posing problems of mixing and pumping operations [28]. As a result, Rico et al. [30] mentioned that low total solids added to long HRT minimizes the need of mixing in anaerobic digestion of dairy manure.

3.4 CONCLUSIONS

Anaerobic digestion of animal manure is looked upon as a strong option in safely reusing wastes or transforming them into valuable materials and energy. This decomposition process that occurs within a confinement contributes to pollution control as it presents with the following benefits; it reduces biological oxygen demand (BOD) and chemical oxygen demand (COD) of wastes; it destroys pathogenic microbes reducing the microbial load to a level which could be safely handled by humans with limited health risks [16] and it destroys volatile fatty acids and many malodorous compounds present in the feedstock and reduces the emission of greenhouse gases. Ultimately, it generates biogas and high quality nutrient-rich fertilizer from animal manure thus upholds the concept of waste to wealth in enhancing sustainability of development [3].

REFERENCES

1. Federal Energy Management Program, Biomass Energy-Focus on Wood Waste. In Biomass and Alternative Methane Fuels; BAMF Fact Sheet: Oak Ridge, TN, USA, 2004.
2. Wilkie, A.C. Biomethane from Biomass, Biowaste and Biofuels. In Bionergy; Wall, J.D., Harwood, C.S., Deamin, A.L., Eds.; ASM Press: Washingston, DC, USA, 2008; pp. 195–215.
3. Uzodinma, E.O.; Ofoefule, A.U.; Eze, J.I.; Mbaeyi, I.; Onwuka, N.D. Effect of some organic wastes on the biogas yield from carbonated soft drink sludge. Sci. Res. Essays 2008, 3, 401–405.
4. Mukumba, P.; Makaka, G.; Mamphweli, S.; Simon, M.; Meyer, E. An insight into the status of biogas digesters technologies in South Africa with reference to the Eastern Cape Province. Fort Hare Pap. 2012, 19, 5–29.
5. Karakashev, D.; Batstone, D.J.; Angelidaki, I. Influence of environmental conditions on methanogenic compositions in anaerobic biogas reactors. Appl. Environ. Microbiol. 2005, 71, 331–338.

6. Kröber, M.; Bekel, T.; Diaz, N.N.; Goesmann, A.; Jaenicke, S.; Krause, L.; Miller, D.; Runte, K.J.; Viehöver, P.; Pühler, A.; et al. Phylognetic characterization of a biogas plant microbial community integrating clone library 16S rDNA sequences and metagenome sequence data obtained by 454-pyrosequencing. J. Biotechnol. 2009, 142, 38–49.

7. Li, J.; Jha, A.K.; He, J.; Ban, Q.; Chang, S.; Wang, P. Assessment of the effects of dry anaerobic co-digestion of cow dung with waste water sludge on biogas yield and biodegradability. Int. J. Phys. Sci. 2011, 6, 3679–3688.

8. Anunputtikul, W.; Rodtong, S. Laboratory Scale Experiments for Biogas Production from Cassava Tubers. In Proceedings of the Joint International Conference on "Sustainable Energy and Environment (SEE)", Hua Hin, Thailand, 1–3 December 2004; 3-017(0). pp. 238–243.

9. Sakar, S.; Yetilmezsoy, K.; Kocak, E. Anaerobic digestion technology in poultry and livestock waste treatment. Waste Manag. Res. 2009, 27, 3–18.

10. St-Pierre, B.; Wright, A.D.G. Metagenomic analysis of methanogen populations in three full- scale mesophilic anaerobic manure digesters operated on dairy farms in Vermont, USA. Bioresour. Technol. 2013, 138, 277–284.

11. Wilkie, A.C. Anaerobic Digestion: Holistic Bioprocessing of Animal Manure. In Proceedings of the Animal Residuals Management Conference, Alexandria, VA, USA, 14–18 October 2000; pp. 1–12.

12. Brown, V.J. Biogas a bright idea for Africa. Environ. Health Perspect. 2006, 114, A300–A303.

13. Liu, F.H.; Wang, S.B.; Zhang, J.S.; Zhang, J.; Yan, X.; Zhou, H.K.; Zhao, G.P.; Zhou, Z.H. The structure of the bacterial and archaeal community in a biogas digester as revealed by denaturing gradient gel electrophoresis and 16S rDNA sequencing analysis. J. Appl. Microbiol. 2009, 106, 952–966.

14. Mauky, E. Biogas Use. Technologies and Trends in Germany; DBFZ, Federal Ministry of Economics and Technology, Eclareon: Berlin, Germany, 2009; pp. 1–28.

15. Rechberger, P. Biogas Markets and Opportunities—A European Review; Anaerobic Digestion in Ireland, Tullamore, AEBIOM: Brussels, Belgium, 2009.

16. Jenkins, S.R.; Armstrong, C.W.; Monti, M.M. Health Effects of Biosolids Applied to Land: Available Scientific Evidence. Virginia Department of Health, 2007. Available online: http://www.vdh.virginia.gov/epidemiology/DEE/documents/biosolids. pdf (accessed on 6 July 2013).

17. Burkholder, J.; Libra, B.; Weyer, P.; Heathcote, S.; Kolpin, D.; Thorne, P.S.; Wichman, M. Impacts of waste from concentrated animal feeding operations on water quality. Environ. Health Perspect. 2007, 115, 308–312.

18. Carbone, S.R.; da Silva, F.M.; Tavares, C.R.G.; Dias Filho, B.P. Bacterial population of a two-phase anaerobic digestion process treating effluent of cassava starch factory. Environ. Technol. 2002, 23, 591–597.

19. Litchfield, J.H. Salmonella Food Poisoning. In Safety of Food, 2nd ed.; Graham, C.W., Ed.; AVI Publishing: Westport, CT, USA, 1980; pp. 120–122.

20. Eriksson, O.; Reich, M.C.; Frostell, B.; Bjorklund, A.; Assefa, G.; Sundqvist, J.-O.; Granath, J.; Baky, A.; Thyselius, L. Municipal solid waste management from a systems perspective. J. Clean. Prod. 2005, 13, 241–252.

21. Willey, J.M.; Sherwood, L.M.; Woolverton, C. Microbial Interactions. In Prescott's Microbiology, 8th ed.; McGraw-Hill Companies Inc.: New York, USA, 2011; pp. 713–728.

22. Nyachoti, C.M.; Omogbenigun, F.O.; Rademacher, M.; Blank, G. Performance responses and indicators of gastrointestinal health in early-weaned pigs fed low-protein amino acid-supplemented diets. J. Anim. Sci. 2006, 84, 125–134.

23. Seppälä, M.; Pyykkönen, V.; Väisänen, A.; Rintala, J. Biomethane production from maize and liquid cow manure-effect of the share of maize, post methanation potential and digestate characteristics. Fuel 2013, 107, 209–216.

24. Nwanta, J.A.; Onunkwo, J.; Ezenduka, E. Analysis of Nsukka metropolitan abattoir solid waste and its bacterial contents in south eastern Nigeria: Public health implication. Arch. Environ. Occup. Health. 2010, 65, 21–26.

25. Ribaudo, M.; Gollehon, N.; Ailley, M.; Kaplan, J.; Johansson, R.; Agapoff, J.; Christenan, L.; Breneman, V.; Peters, M. Manure Management for Water Quality: Costs to Animal Feeding Operations of Applying Manure Nutrients to Land. In Agricultural Economic Report; United States Department of Agriculture: Washington, DC, USA, 2003; No. AER-824; p. 97.

26. Health Care Canada, Guidelines for Canadian Drinking Water Quality: Technical Guideline Document-Bacterial Waterborne Pathogens-Current and Emerging Organisms of Concern. Water Quality and Health Bureau, Healthy Environment and Consumer Safety Branch, Health Canada: Ottawa, ON, Canada, 2006; pp. 1–34.

27. Rapala, J.; Lahti, K.; Rasanen, L.A.; Esala, A.L.; Niemela, S.I.; Sivonen, K. Endotoxins associated with cyanobacteria and their removal during drinking water treatment. Water Res. 2002, 36, 2627–2635.

28. Wilkie, A.C. Anaerobic Digestion of Dairy Manure: Design and Process Considerations. In Dairy Manure Management: Treatment, Handling and Community Relations; Natural Resource, Agriculture, and Engineering Service, Cornell University: Ithaca, NY, USA, 2005; NRAES-176; pp. 301–312.

29. Garcia, M.L.; Angenent, L.T. Interactions between temperature and ammonia in mesophilic digesters for animal waste treatment. Water Res. 2009, 43, 2373–2382.

30. Rico, C.; Rico, J.L.; Muñoz, N.; Gòmez, B.; Tejero, I. Effect of mixing on biogas production during mesophilic anaerobic digestion of screened dairy manure in a pilot plant. Eng. Life Sci. 2011, 11, 476–481.

31. Lutge, B.; Standish, B. Assessing the potential for electricity generation from animal waste biogas on South African farms. Agrekon: Agric. Econ. Res. Policy Pract. S. Afr. 2013, 52, 1–24.

32. Burke, D.A. Dairy Waste Anaerobic Digestion Handbook: Options for Recovering Beneficial Products from Animal Manure; Environmental Energy Company: Olympia, WA, USA, 2001; pp. 1–51. Available online: http://www.makingenergy.com (accessed on 6 July 2013).

33. Tucker, M.F. Farm digesters for small dairies in Vermont. BioCycle 2008, 49, 44.

34. Goodrich, P.R.P.E. Anaerobic Digester Systems for Mid-Sized Dairy Farms; The Minnesota Project: St. Paul, MN, USA, 2005; pp. 1–46.

35. Lozano, C.J.S.; Mendoza, M.V.; de Arango, M.C.; Monroy, E.F.C. Microbiological characterization and specific methanogenic activity of anaerobe sludges used in urban solid waste treatment. Waste Manag. 2009, 29, 704–711.

36. Song, H.; Clarke, W.P.; Blackall, L.L. Concurrent microscopic observations and activity measurements of cellulose hydrolyzing and methanogenic populations during the batch anaerobic digestion of crystalline cellulose. Biotechnol. Bioeng. 2005, 91, 369–378.

37. Franke-Whittle, I.H.; Goberna, M.; Pfister, V.; Insam, H. Design and development of the anaerochip microarray for investigation of methanogenic communities. J. Microbiol. Methods 2009, 79, 279–288.

38. De Graaf, D.; Fendler, R. Biogas Production in Germany; Federal Environment Agency. Dessau Rosslau, Baltic Sea Region Programme. Dessau Rosslau, Baltic Sea Region Programme: Dessau-Rosslau, Germany, 2010; pp. 1–24.

39. Cha, G.-C.; Chung, H.-K.; Kim, D.-J. Characteristics of temperature change on the substrate degradation and bacterial population in one-phase and two-phase anaerobic digestion. Environ. Eng. Res. 2001, 6, 99–108.

40. Demirel, B.; Scherer, P. The roles of acetotrophic and hydrogenotrophic methanogens during anaerobic conversion of biomass to methane: A review. Rev. Environ. Sci. Biotechnol. 2008, 7, 173–190.

41. Brioukhanov, A.L.; Netrusov, A.I.; Eggen, R.I.L. The catalase and superoxide dismutase genes are transcriptionally up-regulated upon oxidative stress in the strictly anaerobic archaeon Methanosarcina barkeri. Microbiology 2006, 152, 1671–1677.

42. Keiner, A.; Leisinger, T. Oxygen sensitivity of methanogenic bacteria. Syst. Appl. Microbiol. 1983, 4, 305–312.

43. Fetzer, S.; Bak, F.; Conrad, R. Sensitivity of methanogenic bacteria from paddy soil to oxygen and desiccation. FEMS Microbiol. Ecol. 1993, 12, 107–115.

44. Anderson, K.L.; Apolinario, E.E.; Sowers, K.R. Desiccation as a long-term survival mechanism for the archaeon Methanosarcina barkeri. Appl. Environ. Microbiol. 2012, 78, 1473–1479.

45. Barber, R.D.; Ferry, J.G. Methanogenesis. Encyclopedia for Life; Nature Publishing Group: New York, NY, USA, 2001; pp. 1–8. Available online: http://www.els.net (accessed on 5 April 2013).

46. McInerney, M.J.; Sieber, J.R.; Gunsalus, R.P. Syntrophy in anaerobic global carbon cycles. Curr. Opin. Biotechnol. 2009, 20, 623–632.

47. Blumer-Schuette, S.E.; Kataeva, I.; Westpheling, J.; Adams, M.W.W.; Kelly, R.M. Extremely thermophilic microorganisms for biomass conversion: status and prospects. Curr. Opin. Biotechnol. 2008, 19, 210–217.

48. Wirth, R.; Kovács, E.; Maròti, G.; Bagi, Z.; Rakhely, G.; Kovács, K.L. Characterization of a biogas—Producing microbial community by short-read next generation DNA sequencing. Biotechnol. Biofuels 2012, 5, 41.

49. Burrell, P.C.; O'Sullivan, C.; Song, H.; Clarke, W.P.; Black-all, L.L. The identification, detection and spatial resolution of Clostridium populations responsible for cellulose degradation in a methanogenic landfill leachate bioreactor. Appl. Environ. Microbiol. 2004, 70, 2414–2419.

50. Li, A.; Chu, Y.; Wang, X.; Ren, L.; Yu, J.; Liu, X.; Yan, J.; Zhang, L.; Wu, S.; Li, S. A pyrosequencing-based metagenomic study of methane-producing microbial community in solid-state biogas reactor. Biotechnol. Biofuels 2013, 6, 3.

51. McInerney, M.J.; Struchtemeyer, C.G.; Sieber, J.; Mouttaki, H.; Stams, A.J.M.; Schnink, B.; Rohlin, L.; Gunsalus, R.P. Physiology, Ecology, Phylogeny, and Ge-

nomics of Microorganisms Capable of Syntrophic Metabolism. Ann. N. Y. Acad. Sci. 2008, 1125, 58–72.

52. Hori, T.; Sasaki, D.; Haruta, S.; Shigematsu, T.; Ueno, Y.; Ishii, M.; Igarashi, Y. Detection of active, potentially acetate-oxidizing syntrophs in an anaerobic digester by flux measurement and formyltetrahydrofolate synthetase expression profiling. Microbiology 2011, 157, 1980–1989.

53. Siriwongrungson, V.; Zeng, R.J.; Angelidaki, I. Homoacetogenesis as the alternative pathway for H2 sink during thermophilic anaerobic degradation of butyrate under suppressed methanogenesis. Water Res. 2007, 41, 4202–4210.

54. Hattori, S.; Galushko, A.S.; Kamagata, Y.; Schink, B. Operation of the CO dehydrogenase/acetyl coenzyme A pathway in both acetate oxidation and formation by the syntrophically acetate oxidizing bacterium Thermacetogenium phaeum. J. Bacteriol. 2005, 187, 3471–3476.

55. Lee, M.J.; Zinder, S.H. Isolation and characterization of a thermophilic bacterium which oxidizes acetate in syntrophic association with a methanogen and which grows acetogenically on H2-CO2. Appl. Environ. Microbiol. 1988, 52, 124–129.

56. Schnürer, A.; Schink, B.; Svensson, B.H. Clostridium ultunense sp. nov., a mesophilic bacterium oxidizing acetate in syntrophic association with a hydrogenotrophic methanogen bacterium. Int. J. Syst. Bacteriol. 1996, 46, 1145–1152.

57. Hattori, S.; Kamagata, Y.; Hanada, S.; Shuon, H. Thermacetogenium phaeum gen. nov., sp. nov., a strictly anaerobic thermophilic, syntrophic acetate-oxidizing bacterium. Int. J. Syst. Evol. Microbiol. 2000, 50, 1601–1609.

58. Balk, M.; Weijma, J.; Stams, A.J.M. Thermotoga lettingae sp. nov., a novel thermophilic, methanol-degrading bacterium isolated from a thermophilic anaerobic reactor. Int. J. Syst. Evol. Microbiol. 2002, 52, 1361–1368.

59. Westerholm, M.; Roos, S.; Schnürer, A. Syntrophaceticus schinkii gen. nov., sp. nov., an anaerobic syntrophic acetate-oxidizing bacterium isolated from a mesophilic anaerobic filter. FEMS Microbiol. Lett. 2010, 309, 100–104.

60. Zhu, W.; Reich, C.I.; Olsen, G.J.; Giometti, C.S.; Yates, J.R. Shotgun proteomics of Methanococcus jannaschii and insights into methanogenesis. J. Proteome Res. 2004, 3, 538–548.

61. Attwood, G.T.; Kelly, W.J.; Altermann, E.H.; Leahy, S.C. Analysis of the Methanobrevibacter ruminantium draft genome: Understanding methanogen biology to inhibit their action in the rumen. Aust. J. Exp. Agric. 2007, 48, 83–88.

62. Ver Eecke, H.C.; Butterfield, D.A.; Huber, J.A.; Lilley, M.D.; Olson, E.J.; Roe, K.K.; Evans, L.J.; Merkel, A.Y.; Cantin, H.V.; Holden, J.F. Hydrogen limited growth of hyperthermophilic methanogens at deep-sea hydrothermal vent. Proc. Natl. Acad. Sci. USA 2012, 109, 13674–13679.

63. Brune, A. Methanogenesis in the Digestive Tracts of Insects. In Handbook of Hydrocarbon and Lipid Microbiology; Timmis, K.W., Ed.; Springer-Verlag: Berlin/Herdelberg, Germany, 2010; pp. 707–728.

64. Westerholm, M.; Levén, L.; Schnürer, A. Bioaugmentation of syntrophic acetate-oxidizing culture in biogas reactors exposed to increasing levels of ammonia. Appl. Environ. Microbiol. 2012, 78, 7619–7625.

65. De Macario, E.C. Taxonomy of Methanogens. In Bergey's Manual of Systematic Bacteriology, 2nd ed.; Springer: New York, NY, USA, 2008.

66. Batstone, D.J.; Keller, J.; Angelidaki, I.; Kalyuzhnyi, S.V.; Pavlostathis, S.G.; Rozzi, A.; Sanders, W.T.M.; Siegrist, H.; Vavilin, V.A. The IWA anaerobic digestion model No.1 (ADM1). Water Sci. Technol. 2002, 45, 65–73.

67. Krakat, N.; Westphal, A.; Schmidt, S.; Scherer, P. Anaerobic digestion of renewable biomass: Thermophilic temperature governs methanogen population dynamics. Appl. Environ. Microbiol. 2010, 76, 1842–1850.

68. Krakat, N.; Schmidt, S.; Scherer, P. Mesophilic fermentation of renewable biomass: Does hydraulic retention time regulate methanogen diversity. Appl. Environ. Microbiol 2010, 76, 6322–6326.

69. Klocke, M.; Mähnert, P.; Mundt, K.; Souidi, K.; Linke, B. Microbial community analysis of a biogas-producing completely stirred tank reactor fed continuously with fodder beet silage as mono-substrate. Syst. Appl. Microbiol. 2007, 30, 139–151.

70. Klocke, M.; Nettmann, E.; Bergmann, I.; Mundt, K.; Souidi, K.; Mumme, J.; Linke, B. Characterization of the methanogenic archaea within two-phase biogas reactor systems operated with plant biomass. Syst. Appl. Microbiol. 2008, 31, 190–205.

71. Amani, T.; Nosrati, M.; Sreekrishnan, T.R. Anaerobic digestion from the viewpoint of microbiological, chemical, and operational aspects: A review. Environ. Rev. 2010, 18, 255–278.

72. Solera, R.; Romero, L.I.; Sales, D. Determination of the microbial population in thermophilic anaerobic reactor: Comparative analysis by different counting methods. Anaerobe 2001, 7, 79–86.

73. Ziganshin, A.M.; Schmidt, T.; Scholwin, F.; Il'inskaya, O.N.; Harms, H.; Kleinsteuber, S. Bacteria and Archaea involved in anaerobic digestion of distillers grains with solubles. Appl. Microbiol. Biotechnol. 2011, 89, 2039–2052.

74. Rivière, D.; Desvignes, V.; Pelletier, E.; Chaussonnerie, S.; Guermazi, S.; Weissenbach, J.; Li, T.; Camacho, P.; Sghir, A. Towards the definition of a core of microorganisms involved in anaerobic digestion of sludge. Int. Soc. Microb. Ecol. 2009, 3, 700–714.

75. Scully, C.; Collins, G.; O'Flaherty, V. Assessment of anaerobic wastewater treatment failure using terminal restriction fragment length polymorphism analysis. J. Appl. Microbiol. 2005, 99, 1463–1471.

76. Kataoka, N.; Tokiwa, Y.; Takeda, K. Improved technique for identification and enumeration of methanogenic bacterial colonies on roll tubes by epifluorescence microscopy. Appl. Environ. Microbiol. 1991, 57, 3671–3673.

77. Singh, L.S.; Mazumder, P.B. Differential approaches for studying methanogens: Methods, analysis and prospects. Assam Univ. J. Sci. Technol. 2010, 6, 123–128.

78. Schlüter, A.; Bekel, T.; Diaz, N.N.; Dondrup, M.; Eichenlaub, R.; Gartemann, K-H.; Krahn, I.; Krause, L.; Krömeke, H.; Kruse, O.; et al. The metagenome of a biogas-producing microbial community of a production-scale biogas plant fermenter analyzed by the 454-pyrosequencing technology. J. Biotechnol. 2008, 136, 77–90.

79. Lee, C.; Kim, J.; Shin, S.G.; Hwang, S. Monitoring bacterial and archaeal community shifts in a mesophilic anaerobic batch reactor treating a high-strength organic wastewater. FEMS Microbiol. Ecol. 2008, 65, 544–554.

80. Cirne, D.G.; Lehtomäki, A.; Björnsson, L.; Blackall, L.L. Hydrolysis and microbial community analyses in two-stage anaerobic digestion of energy crops. J. Appl. Microbiol. 2006, 103, 516–527.

81. Jaenicke, S.; Ander, C.; Bekel, T.; Bisdorf, R.; Dröge, M.; Gartemann, K.-H.; Jüneman, S.; Kaiser, O.; Krause, L.; Tille, F.; et al. Comparative and joint analyses of two-metagenomic dataset from a biogas fermenter obtained by 454-pyrosequencing. PLoS One 2011, 6, e14519.

82. Li, M.; Cao, H.; Hong, Y.-G.; Gu, J.-D. Seasonal dynamics of anammox bacteria in estuarial sediment Mai Po nature reserve revealed by analyzing the 16S rRNA and hydrazine oxidoreductase (hzo) genes. Microbes Environ. 2011, 26, 15–22.

83. Ozgun, D.; Basak, S.; Cinar, O. Current Molecular Biologic Techniques for Anaerobic Ammonium Oxidizing (Anammox) Bacteria. In Proceedings of the Sixteenth International Water Technology Conference, IWTC, Istanbul, Turkey, 7–10 May 2012; pp. 1–15.

84. Harhangi, H.R.; Roy, M.L.; Alen, T.V.; Hu, B.-L.; Groen, J.; Kartal, B.; Tringe, S.G.; Quan, Z.-X.; Jetten, M.S.M.; den Camp, H.J.M.O. Hydrazine synthase, a unique phylomarker with which to study the presence and biodiversity of anammox bacteria. Appl. Environ. Microbiol. 2012, 78, 752–758.

85. Zhou, M.; McAllister, T.A.; Guan, L.L. Molecular identification of rumen methanogen: Technologies, advances and prospects. Anim. Feed Sci. Technol. 2011, 166–167, 76–86.

86. Lutton, P.E.; Wayne, J.M.; Sharp, R.J.; Riley, P.W. The mcrA gene as an alternative to 16S rRNA in the phylogenetic analysis of methanogen population in landfills. Microbiology 2002, 148, 3521–3530.

87. Denman, S.E.; Tomkins, N.W.; McSweeney, C.S. Quantitation and diversity analysis of ruminal methanogenic populations in response to the antimethanogenic compound bromochloromethane. FEMS Microbiol. Ecol. 2006, 62, 313–322.

88. Jiang, B.; Song, K.; Ren, J.; Deng, M.; Sun, F.; Zhang, X. Comparison of metagenomic samples using sequence signatures. BMC Genomics 2012, 13, 730.

89. Mashhadi, Z. Identification and Characterization of the Enzymes Involved in Biosynthesis of FAD and Tetrahydromethanopterin in Methanococcus jannaschii. Ph.D. Thesis, Virginia Polytechnic Institute and State University (Virginia Tech), Blacksburg, VA, USA, In: Doctor of philosophy Biochemistry.. 30 June 2010; pp. 1–136.

90. Leahy, S.C.; Kelly, W.J.; Altermann, E.; Ronimus, R.S.; Yeoman, C.J.; Pacheco, D.M.; Li, D.; Kong, Z.; McTavish, S.; Sang, C.; et al. The genome sequence of the rumen methanogen Methanobrevibacter ruminantium reveals new possibilities for controlling ruminant methane emission. PLoS One 2010, 5, e8926.

91. Handelsman, J. Metagenomics: Application of genomics to uncultured microorganisms. Microbiol. Mol. Biol. Rev. 2004, 68, 669–685.

92. Gilbert, J.A.; Dupont, C.L. Microbial metagenomics: Beyond the genome. Annu. Rev. Mar. Sci. 2004, 3, 347–371.

93. Ilaboya, I.R.; Assekhame, F.F.; Ezugwu, M.O.; Erameh, A.A.; Omofuma, F.E. Studies on biogas generation from agricultural wastes; analysis of the effects of alkaline on gas generation. World Appl. Sci. J. 2010, 9, 537–545.

94. Umaña, O.; Nikolaeva, S.; Sanchez, E.; Borja, R.; Raposo, F. Treatment of screened dairy manure by upflow anaerobic fixed bed reactors packed with waste tyre rubber and a combination of waste tyre rubber and Zeolite: Effect of the hydraulic retention time. Bioresour. Technol. 2008, 99, 7412–7417.

95. Balsam, J.; Ryan, D. Anaerobic Digestion of Animal Wastes: Factors to Consider; ATTRA: Butte, MT, USA, 2006; pp. 1–10.

96. Cioabla, A.E.; Lonel, L.; Dumitrel, G.-A.; Popescu, F. Comparative study on factors affecting anaerobic digestion of agricultural vegetal residues. Biotechnol. Biofuels 2012, 5, 39.

97. Choorit, W.; Wisarnwan, P. Effect of temperature on the anaerobic digestion of palm oil mill effluent. Electron. J. Biotechnol. 2007, 10, 376–385.

98. Saleh, M.M.A.; Mahmood, U.F. Anaerobic Digestion Technology for Industrial Waste Water Treatment. In Proceedings of the Eighth International Water Technology Conference, IWTC, Alexandria, Egypt, 26–28 March 2004; pp. 817–833.

99. Rittmann, B.E.; McCarty, P.L. Environmental Biotechnology: Principles and Applications; McGraw-Hill Book Co.: New York, NY, USA, 2001; p. 768.

100. El-Mashad, H.M.; Zeeman, G.; van Loon, W.K.P.; Bot, G.P.A.; Lettinga, G. Effect of temperature and temperature fluctuation on thermophilic anaerobic digestion of cattle manure. Bioresour. Technol. 2004, 95, 191–201.

101. Campos, E.; Palatsi, J.; Flotats, X. Co-Digestion of Pig Slurry and Organic Wastes from Food Industry. In Proceedings of the 2th International Symposium on Anaerobic Digestion of Solid Waste, Barcelona, Junio, Spain, 15–18 June 1999; pp. 192–195.

102. Gerardi, M.H. The Microbiology of Anaerobic Digesters; John Wiley and Sons Inc.: Hoboken, NJ, USA, 2003; pp. 91–118.

103. Molinuevo-Salces, B.; García-González, M.C.; González-Fernández, C.; Cuetos, M.J.; Morán, A.; Gòmez, X. Anaerobic co-digestion of livestock wastes with vegetable processing wastes: A statistical analysis. Bioresour. Technol. 2010, 101, 9479–9485.

104. Veeken, A.; Kalyuzhnyi, S.; Scharff, H.; Hamelers, B. Effect of pH and VFA on hydrolysis of organic solid waste. J. Environ. Eng. 2000, 126, 1076–1081.

105. El Hadj, T.B.; Astals, S.; Galí, A.; Mace, S.; Mata-Álvarez, J. Ammonia influence in anaerobic digestion of OFMSM. Water Sci. Technol. 2009, 59, 1153–1158.

106. Chen, Y.; Cheng, J.J.; Creamer, K.S. Inhibition of anaerobic digestion process. A review. bioresour. Technol. 2008, 99, 4044–4064.

107. Strik, D.P.B.T.B.; Domnanovich, A.M.; Holubar, P. A pH-based control of ammonia in biogas during anaerobic digestion of artificial pig manure and maize silage. Process Biochem. 2006, 41, 1235–1238.

108. Angelidaki, I.; Ahring, B.K. Thermophilic anaerobic digestion of livestock waste: The effect of ammonia. Appl. Microbiol. Biotechnol. 1993, 38, 560–564.

109. Bolzonella, D.; Pavan, P.; Battistoni, P.; Cecchi, F. Mesophilic anaerobic digestion of waste activated sludge: Influence of solid retention time in the wastewater treatment process. Process Biochem. 2005, 40, 1453–1460.

110. Karim, K.; Hoffmann, R.; Klasson, K.T.; Al-Dahhan, M.H. Anaerobic digestion of animal waste: Effects of mode of mixing. Water Res. 2009, 39, 3597–3606.

111. Babaee, A.; Shayegan, J. Effects of Organic Loading Rates (OLR) on Production of Methane from Anaerobic Digestion of Vegetable Waste. In Proceedings of the World Renewable Energy Congress, Linköping, Sweden, 8–13 May 2011; pp. 411–417.

112. Rincòn, B.; Travieso, L.; Sanchez, E.; Martín, M.L.A.; Martín, A.; Raposo, F.; Borja, R. The effect of organic loading rate on the anaerobic digestion of two-phase olive

mill solid residue derived from fruits with low ripening index. J. Chem. Technol. Biotechnol. 2007, 82, 259–266.

113. Rincòn, B.; Borja, R.; González, J.M.; Portillo, M.C.; Sáiz-Jiménez, C. Influence of organic loading rate and hydraulic retention time on the performance, stability, and microbial communities of one-stage anaerobic digestion of two-phase olive mill solid residue. Biochem. Eng. J. 2008, 40, 253–261.

114. Tchobanoglous, G.; Burton, F.L.; Stensel, H.D. Wastewater Engineering: Treatment and Reuse, 4th ed.; Metcalf and Eddy, Inc., Tata Mcgraw-Hill Publishing Company Ltd.: New Delhi, India, 2003.

115. Zhang, R.; El-Mashad, H.M.; Hartman, K.; Wang, F.; Liu, G.; Choate, C.; Gamble, P. Characterization of food waste as feedstock for anaerobic digestion. Bioresour. Technol. 2007, 98, 929–935.

116. Matseh, I. Effect of Ni and Co as trace elements on digestion performance and biogas produced from the fermentation of palm oil mill effluent. Int. J. Waste Resour. 2012, 2, 16–19.

117. Gustavsson, J. Cobalt and Nickel Bioavailability for Biogas Formation. Ph.D. Thesis, Department of Thematic studies, University of Linköping, Linköping, Sweden, Water and Environmental Studies. 19 January 2012; pp. 1–64.

118. Pobeheim, H.; Munk, B.; Johansson, J.; Guebitz, G.M. Influence of trace elements on methane formation froma synthetic model substrate for maize silage. Bioresour. Technol. 2010, 101, 836–839.

119. Facchin, V.; Cavinato, C.; Fatone, F.; Pavan, P.; Cecchi, F.; Bolzonella, D. Effect of trace element supplementation on the mesophilic anaerobic digestion of food wastes in batch trials. The influence of inoculum origin. Biochem. Eng. J. 2013, 70, 71–77.

120. Rojas, C.; Fang, S.; Uhlenhut, F.; Borchert, A.; Stein, I.; Schlaak, M. Stirring and biomass starter influences the anaerobic digestion of different substrates for biogas production. Eng. Life Sci. 2010, 10, 339–347.

121. Ghanimeh, S.; El Fadel, M.; Saikaly, P. Mixing effect on thermophilic anaerobic digestion of source-sorted organic fraction of municipal solid waste. Bioresour. Technol. 2012, 117, 63–71.

CHAPTER 4

New Steady-State Microbial Community Compositions and Process Performances in Biogas Reactors Induced by Temperature Disturbances

GANG LUO, DAVIDE DE FRANCISCI, PANAGIOTIS G. KOUGIAS, TREU LAURA, XINYU ZHU, AND IRINI ANGELIDAKI

4.1 BACKGROUND

Anaerobic digestion (AD) is widely used in the treatment of organic wastes to achieve reduction of the wastes with simultaneous production of biogas [1]. The production of biogas via AD is a complex process, involving many different microbial species [2]. The complex organic compounds are first hydrolyzed into oligomers and monomers, and then further converted to acetate, CO_2, and H_2 by various fermenting bacteria. The methanogenesis is the final step to convert acetate, CO_2, and H_2 to CH_4 by methanogenic archaea. The syntrophic relationship between bacteria and archaea is essential for the stability of the biogas process [3].

It is crucial to understand the processes and factors controlling the microbial community composition in biogas reactors. The mechanisms of community assembly and the critical factors shaping species composition and structure remain controversial in ecology [4]-[6]. The traditional niche-based theory supports the idea that the community is shaped mainly by deterministic factors such as competition and niche differentiation, and thereby asserts that community composition should converge toward a single pattern under similar environmental conditions [7]. In contrast to niche-based theory, neutral theory assumes that many natural community patterns can be generated under similar environmental conditions by stochastic factors considering birth, death, dispersal, and speciation and disregards the differences between species at the same trophic level [8]. In addition, disturbance was also shown to play an important role in the community assembly since the disturbance could kill or damage certain species and promote the growth of other species that are resistant to the disturbance [9],[10].

Microbial community compositions in biogas reactors have been studied for several decades, and deterministic factors including temperature, hydraulic retention time (HRT), and substrate type have been demonstrated to play an important role in shaping microbial communities [11],[12]. However, based on the neutral theory, stochastic factors may be important in shaping the highly diverse microbial communities in biogas reactors. Up to now, it is still unknown whether there are different microbial community patterns under the same environmental conditions in biogas reactors if stochastic factors are determining the microbial communities. Although different disturbances (such as temperature and organic loading) on the biogas process have been evaluated before, most of the studies focused only on the reactor performances, and the effect of disturbance on the community assembly was not documented [13]-[16]. It is still unknown whether the disturbance in the biogas reactors would lead to different steady-state microbial community compositions and reactor performances.

Traditional molecular technologies for microbial community analysis (for example, polymerase chain reaction-denaturing gradient gel electrophoresis, terminal restriction fragment length polymorphism, and cloning) can only identify the most abundant microorganisms in the microbial community. The less abundant yet functionally important microorganisms cannot be detected [17]. Therefore, using traditional molecular technologies, it is difficult to study variations of less abundant microorganisms in different samples. With the newly developed sequencing technologies, it is possible to define microbial community composition with a high sequencing depth. The Ion Torrent Personal Genome Machine (PGM) (Life Technologies), launched in early 2011, has the highest throughput compared with 454 GS Junior (Roche) and Miseq (Illumina), thus making the high-throughput sequencing cost effective and time saving [18].

Based on the above considerations, the objective of this study was to understand the role of stochastic factors (based on neutral theory) and disturbance in the steady-state microbial community assembly and functions in biogas reactors. We ran three replicate biogas reactors treating cattle manure to first determine whether similar microbial communities would be achieved at steady states where the reactors were operated under the same conditions. In most modern biogas plants one attempts to keep a constant temperature, as temperature stability is of utmost importance for the biogas process. Nevertheless, biogas reactors may be subjected to undesired temperature fluctuations due to various technical problems such as heat exchanger or pump failures or fouling in temperature sensors [16],[19]. Therefore, temperature disturbance was introduced to the three reactors in order to determine if and to what extent the temperature disturbance would alter the steady-state microbial community. In addition, the reactor performances including biogas production, pH, and total volatile fatty acids (VFAs) were monitored. The microbial community composition was analyzed by Ion Torrent sequencing of 16 s rRNA gene amplicons.

4.2 RESULTS AND DISCUSSION

4.2.1 REACTOR PERFORMANCE

The monitoring profiles for methane yield, pH, and total VFAs in the three reactors are shown in Figure 1, and the overall performances of the reactors at steady state are summarized in Table 1. The initial higher methane yield was due to the methane production from the inoculum, since there are still organics in the inoculum which can be digested. After around 30 days' operation, the methane yields were relatively stable. The steady-state methane yields for the three reactors were not significantly different (P<0.05): 194 ± 7.3, 189 ± 14.5, and 195 ± 6.9 mL/g volatile solids (VS) for reactors A, B, and C, respectively. Both the pH values (around 7.5) and the total VFAs concentrations were also similar for all three reactors. Acetate was the dominant VFAs, as seen in Figure 2. The above results indicated that the replicate reactors (A, B, and C) did not present obvious differences in their performances.

TABLE 1: Summary of the reactor performances at steady state before (phase I, 0 to 50 days) and after (phase II, 61 to 112 days) temperature disturbance

Parameter	A_b	A_a	B_b	B_a	C_b	C_a
Methane yield (mL CH_4/gVS)	194 ± 7.3	220 ± 17.5	189 ± 14.5	213 ± 11.7	195 ± 6.9	214 ± 13.1
pH	7.50 ± 0.06	7.64 ± 0.02	7.51 ± 0.04	7.66 ± 0.01	7.52 ± 0.04	7.60 ± 0.02
Total VFAs (mM)	25.5 ± 3.9	3.3 ± 0.9	24.9 ± 1.7	5.3 ± 0.6	21.1 ± 1.9	7.8 ± 1.2

Subscript b means before temperature disturbance, subscript a means after temperature disturbance.

From day 50 (phase II), the temperatures of the reactors were changed (A 25°C, B 45°C, C 55°C) from the original temperature of 37°C. A sharp decrease in methane yields was observed in all the reactors, together with a decrease in pH and an increase in total VFAs. After 10 days at the new temperatures, the total VFAs increased to around 60 mM for reactor A, and to around 90 mM for both reactors B and C, which clearly indicates that the increase of temperature had a more profound effect on the stability of the reactors. There are several reasons leading to the higher VFAs accumulation when the temperatures were increased. It could be due to faster adaptation of acidogenic bacteria or to a greater temperature span of acidogens at higher temperatures compared to methanogens, as they generally grow more slowly and have a narrower temperature span [20],[21]. Among the total VFAs, acetate was still the dominant component, although propionate was also accumulated, which is consistent with previous reports that propionate is easy to accumulate when the biogas reactor is disturbed [22],[23].

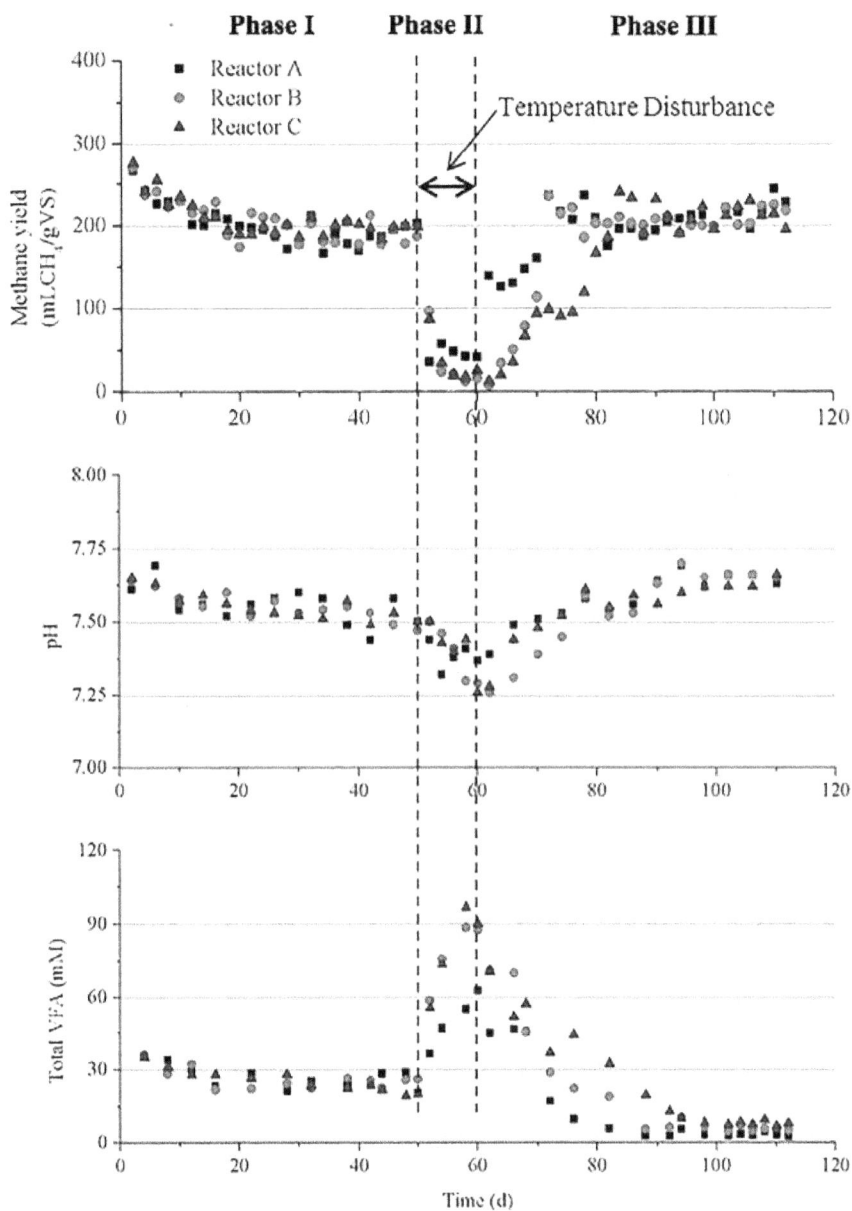

FIGURE 1: Reactor performances for the whole operational period.

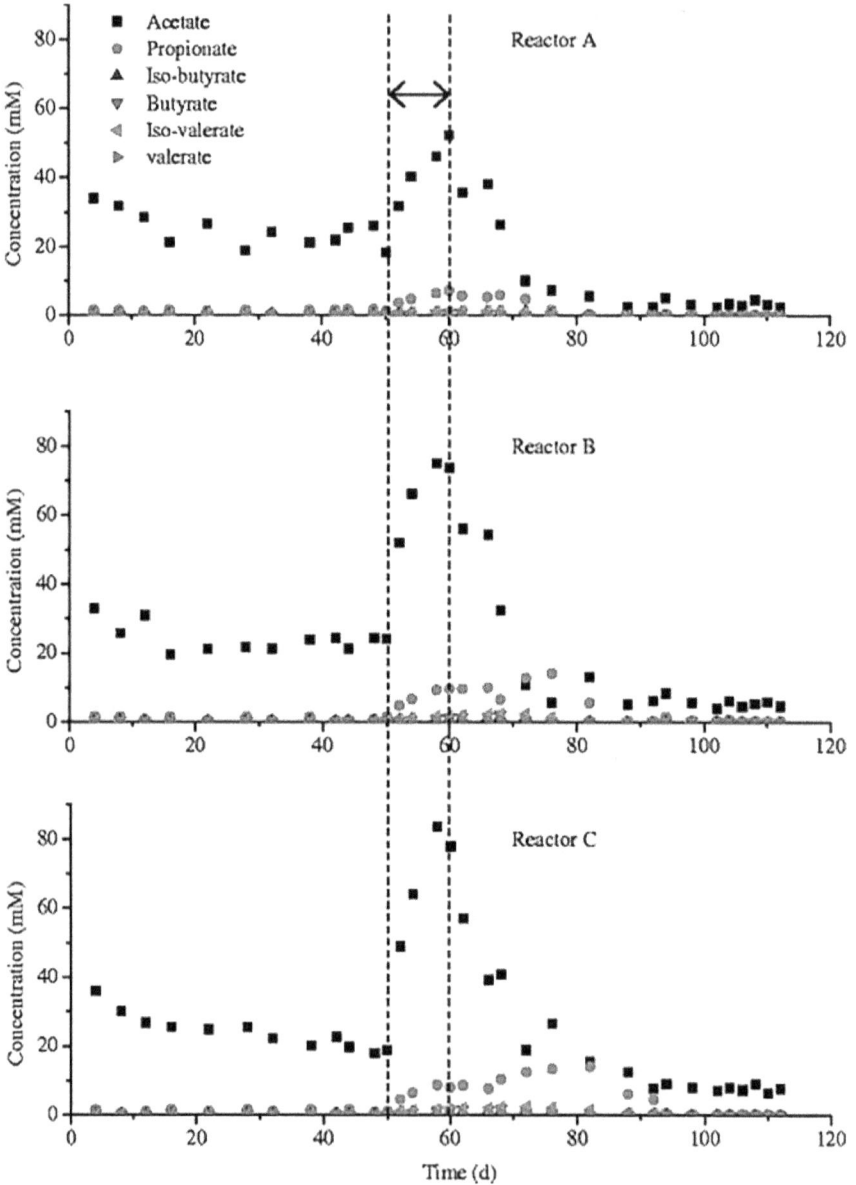

FIGURE 2: Individual VFAs changes for the whole operational period.

From day 60 (phase III), the operating temperatures of all the reactors were changed back to 37°C. The methane yield increased immediately for reactor A. However, a slower recovery of the methane yield was observed for reactors B and C, which might indicate that the higher temperature disturbances (45°C and 55°C) had a more negative impact on the stability of the biogas reactors. In particular, reactor C took around 10 days before the methane yield increased to a similar level to that of the steady-state level in phase I. The fast increase of methane yield in reactor A was in good agreement with the fast decrease of total VFAs. During the steady states of phase III, the methane yields for reactors A, B, and C were 220 ± 17.5, 213 ± 11.7, and 214 ± 13.1 mL/gVS, respectively, and there were no significant differences ($P < 0.05$) among the reactors. However, the methane yields were all significantly higher (around 10%) than those in phase I. The results indicated that the temperature disturbances affected the performances of all the reactors, although the operational conditions were the same for all the reactors in both phases I and III. The lower VFAs concentrations in phase III compared with those in phase I could explain the increased methane yields. However, the decreased VFAs concentration only accounted for 10% or less of the increased methane yield. This indicates that the increased methane yield could also be related with the increased hydrolysis of the solid part in cattle manure, which would lead to higher methane production. The lower VFAs concentrations in phase III for all the three reactors also resulted in the relatively higher pH (around 7.6).

4.2.2 MICROBIAL COMMUNITY ANALYSIS

The numbers of sequences after quality filtration from different samples are shown in Additional file 1: Table S1. The average sequence lengths were around 273 bp for all the samples. The high-quality sequences were assigned to taxonomic classifications by the Ribosomal Database Project (RDP) classifier. Since the primers used in the present study were universal primers, sequences belonging to both bacteria and archaea were obtained at the same time [24],[25]. The phylogenetic classification of sequences assigned to bacteria from all the samples is summarized in Figure 3. Samples A1, B1, and C1 had similar distributions of the sequences at the phylum, class, and genus levels. *Firmicutes, Bacteroidetes*, and *Proteobacteria* were dominant at the phylum level, and their dominance in biogas reactors was in accordance with other studies [26],[27]. *Clostridia, Bacteroidia*, and *Gammaproteobacteria* were dominant at the class level. However, a considerable amount of the sequences (around 50%) were unclassified at the genus level, which could be due to some new microorganisms that have not yet been identified. The high percentages of unclassified sequences at the genus level were also found in previous studies [1],[28]. The temperature disturbances had different effects on the shift of bacterial communities in reactors A, B, and C. The decrease of temperature from 37°C to 25°C in reactor A resulted in an increased abundance of *Proteobacteria* (A2), and the reason might be that some of the classes (for example, *Gammaproteobacteria*) belonging to *Proteobacteria* can grow well at lower temperatures [29]. In reactor B, the increase of temperature from 37°C to 45°C led to an increased relative abundance of the unclassified sequences at the phylum level (B2). The increased relative abundance of *Firmicutes* (C2) was observed with a further increase of temperature from 37°C to 55°C in reactor C. The relative abundance of class *Clostridia*, belonging to *Firmicutes*, was enriched in sample C2, which could be due to their spore-forming character and gradual adaptation to thermophilic conditions [26],[30]. The bacterial communities (A3, B3, and C3, sampled on day 60) continued to change after the temperatures in all the reactors returned to 37°C, which was consistent with the unstable reactor performances (Figure 1, day 60). Samples A4, B4, and C4 were obtained during the steady states of phase III, and they had similar distributions at the phylum, class, and genus levels. However, compared with A1, B1, and C1, the relative abundance of *Bacteroidetes* in A4, B4, and C4 increased and that of *Proteobacteria* decreased. Differences at the class and genus levels were also observed between A1, B1, C1 and A4, B4, C4. The above results showed that although the biogas reactors, before and after the temperature disturbance, were run under exactly the same operational conditions, the bacterial communities did not return to the original bacterial composition. New steady-state bacterial community compositions, distinct from the original, were established after the temperature disturbances.

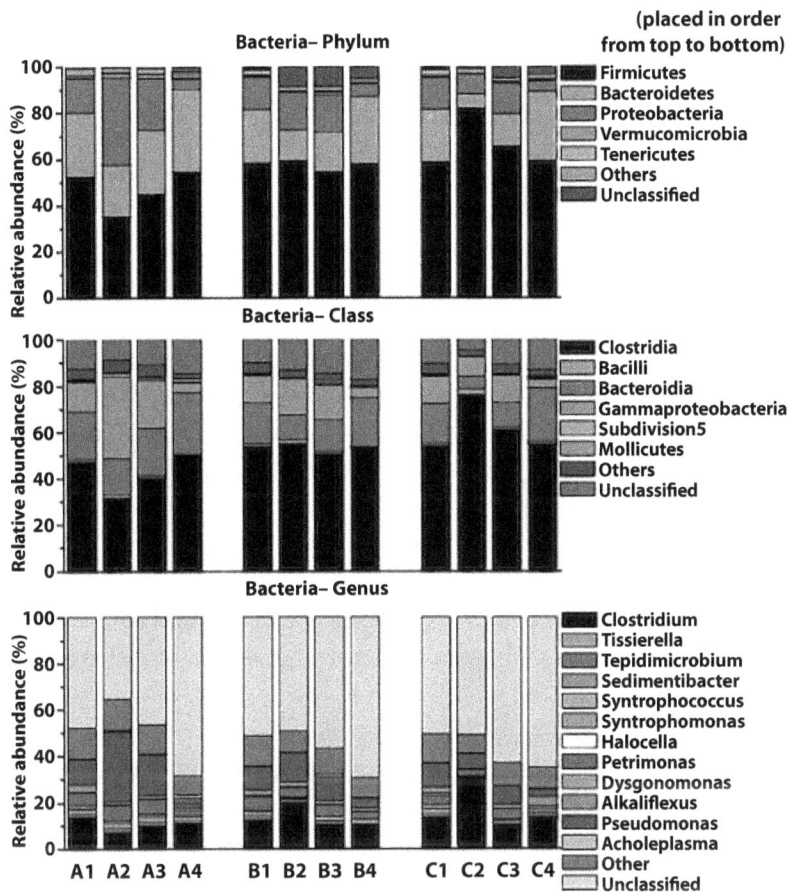

FIGURE 3: Taxonomic classification of the bacteria communities. Phyla, classes, and genera making up less than 1% of total composition in all the samples were classified as Others.

The phylogenetic classification of sequences assigned to archaea from all the samples is summarized in Figure 4. The archaea mediating hydrogenotrophic and aceticlastic methanogenesis were found mainly within four orders (*Methanobacteriales*, *Methanococcales*, *Methanomicrobiales*, and *Methanosarcinales*) [31]. Therefore, only order- and genus-level classifications are shown in Figure 4. Samples A1, B1, and C1 were all dominated by *Methanomicrobiales* and *Methanobacteriales* in phase I, which belonged to hydrogenotrophic methanogens. Similar distributions of samples A1, B1, and C1 at the genus level were also observed, and the dominant genera were *Methanoculleus*, *Methanocorpusculum*, *Methanobrevibacter*, and *Methanobacterium*. An increase of *Methanobacteriales* was found in all the reactors after temperature disturbance, which may indicate that the archaea belonging to this order were more resistant to the temperature disturbance (both downwards and upwards). It has been reported that the most frequently observed hydrogen utilizers are members of *Methanobacteriales*, present in both manure and sewage sludge digesters [31]. Further study is needed in order to understand why *Methanobacteriales* were more resistant to temperature changes than other methanogens. In reactor C, the temperature increase resulted in the increased relative abundance of order *Methanosarcinales*, which are mainly aceticlastic methanogens. Since the methane production during temperature shock was significantly reduced, the changes of archaeal communities during the temperature disturbance might be due to the different decay rates of the archaea rather than the different growth rates of the archaea. After the temperature was changed back to 37°C, *Methanosarcinales* became dominant in all the reactors (A3, B3, and C3). At the steady states of phase III, *Methanomicrobiales*, *Methanosarcinales*, and *Methanobacteriales* were all dominant in reactors A, B, and C. The genus *Methanosarcina*, belonging to *Methanosarcinales*, mainly mediates aceticlastic methanogenesis, and therefore it is expected that the methanogenic pathway was changed before and after temperature disturbance. The dominance of *Methanosarcina* might be related to the better reactor performances in phase III compared with phase I. It has been reported that the dominance of hydrogenotrophic methanogenesis is always related to extreme conditions such as high ammonia or acetate concentration [32]-[34]. It is possible that the higher acetate concentration in phase I in all the reactors induced the dominance of archaea belonging to hydrogenotrophic methanogens.

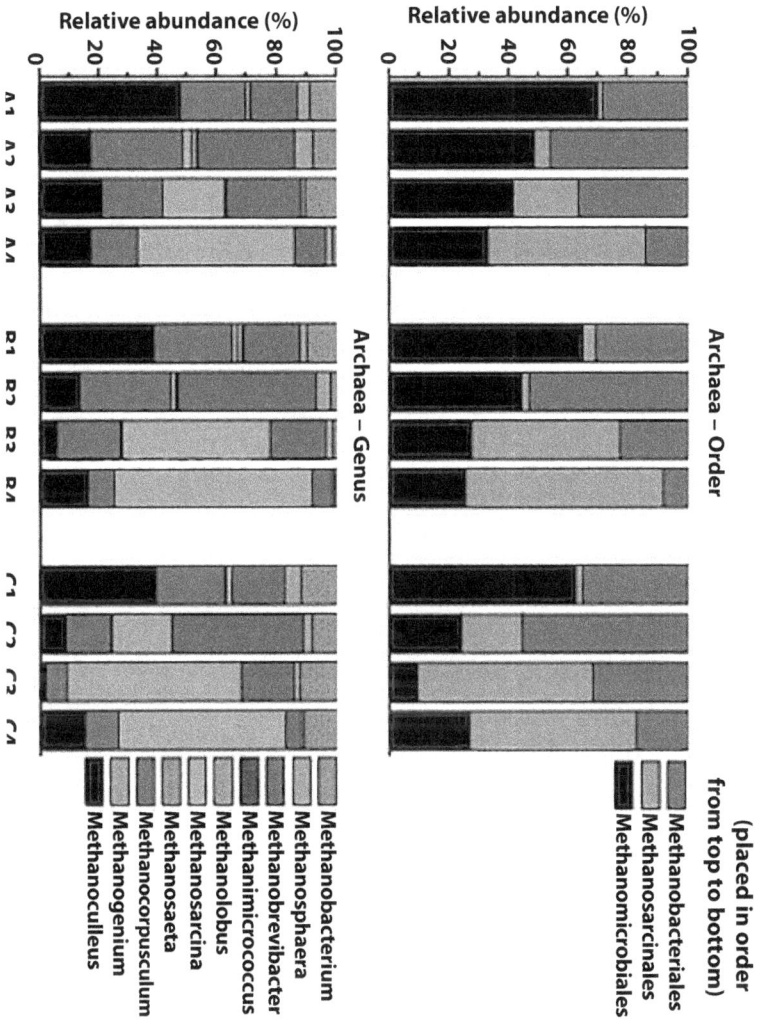

FIGURE 4: Taxonomic classification of the archaea communities.

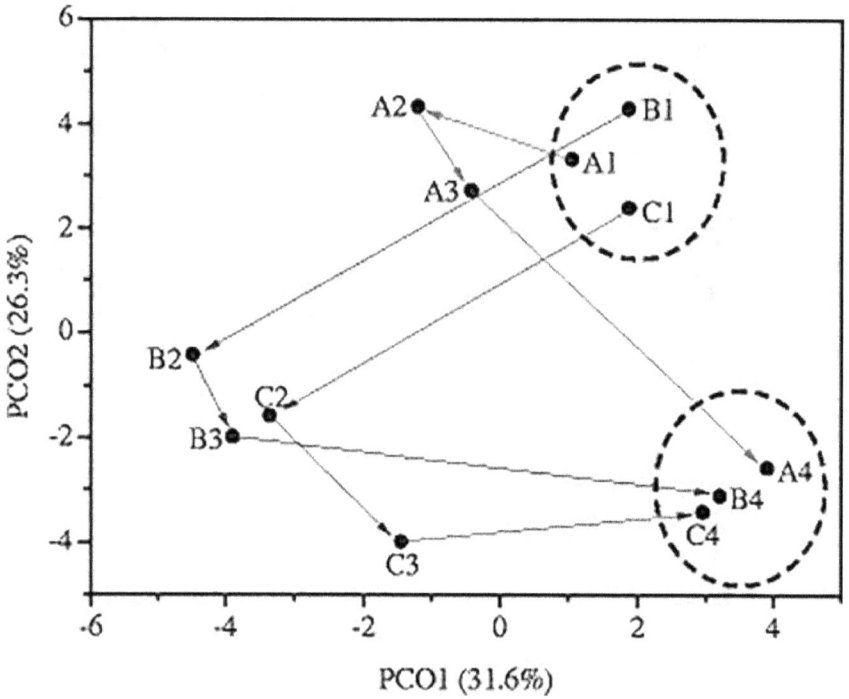

FIGURE 5: PCoA of all the samples.

The differences between the microbial communities of different samples were further assessed by principal coordinates analysis (PCoA) and hierarchical cluster analysis. The results from PCoA are shown in Figure 5. Principal components 1 and 2 explained 31.6% and 26.3% of the total community variations, respectively. A1, B1, and C1 were clustered together, and were well separated from the group of A4, B4, and C4. The results were consistent with the taxonomic analysis that steady-state microbial community compositions were changed after temperature disturbances in reactors A, B, and C, which could also be used to explain the different steady-state reactor performances in phase I and phase III, as discussed in the preceding section. A2 was close to A1, while B2 and C2 were far away from B1 and C1, which indicated that the increase of temperature to 45°C or 55°C in the biogas reactor induced significant changes in the microbial communities. A3, B3, and C3 were closer to A4, B4, and C4, which suggested that the microbial communities gradually changed to the new steady states after the temperature changed back to 37°C. The hierarchical cluster analysis (Additional file 1: Figure S1) also clustered A1, B1, and C1 as one group and A4, B4, and C4 as one group, which further supported the cluster results from PCoA.

4.2.3 MICROBIAL COMMUNITY ASSEMBLY AND FUNCTION IN ANAEROBIC DIGESTION OF CATTLE MANURE

The results in the present study clearly showed that replicate biogas reactors treating cattle manure had similar steady-state reactor performances under the same environmental conditions. The replicate biogas reactors also had similar microbial community compositions based on RDP classification, PCoA, and hierarchical cluster analysis. However, Zhou et al. found that, under the same environmental conditions, both bacterial community compositions and functions in replicate microbial electrolysis cells were different, and they proposed that stochastic assembly played a dominant role in determining not only community structure but also ecosystem functions [35]. Based on our results, the microbial community compositions and functions in anaerobic digestion of cattle manure were not obviously affected by stochastic assembly. It seems that stochastic assembly played different roles in different ecosystems. A recent publication, which found that community history affects the predictability of microbial ecosystem development, might explain the differences between our results and Zhou's results. Pagaling et al.[36] demonstrated that the final community composition and function are unpredictable when the source communities (inoculum) colonize a novel environment, while the community development is more reproducible when source communities are pre-conditioned to their new habitat. In our study, the inoculum was obtained from a biogas reactor, which was adapted to the anaerobic condition for methane production. Therefore, the steady-state microbial community composition and reactor performance were reproducible in the replicate biogas reactors. However, all the reactors in Zhou's study were inoculated with wastewater not pre-conditioned in the microbial electrolysis cell reactors, which might have been the reason for the variation of microbial community compositions and functions observed. Further investigation is imperative to elucidate whether other inocula sources (sewage sludge, soil, and others), which are not derived from anaerobic reactors, would lead to different stable-state microbial communities and reactor performances in biogas reactors under the same operational conditions.

Although the effect of temperature disturbance on biogas production has been studied before [13],[16], the previous studies only focused on the reactor performances. The rapid recovery of biogas production after the temperature disturbance was also previously observed [13], but in the cited study the biogas production reached the same level as that prior to the disturbance. We are the first to report the increased biogas production after temperature disturbance. The difference in biogas production recovery levels could be due to the fact that the substrate was already efficiently degraded before temperature shock in Chae's study [13]. Our results suggested that temperature disturbance could be used as a strategy to stimulate biogas production in the biogas reactor, where the biogas production was not so efficient (possibly due to lower HRT). In a previous study, Nielsen et al. investigated the effects of disturbance of oleate on the performance of a biogas reactor treating cattle and pig manure, and they also found that a lower VFAs concentration along with a higher methane production were achieved in the biogas reactor after the disturbance of oleate

[15]. However, they could not link the microbial community composition with the reactor performance due to the lack of microbial analysis. Our results combined with the above literature suggest that the idea that different types (temperature, oleate) of disturbances might have similar stimulation effects on the biogas process, and the effects of other disturbances (ammonia, organic loading shock, and others) on the community assembly and functioning of the biogas reactor deserve to be further investigated. Although in the present study, three different temperature disturbances were investigated, the establishment of new steady-state microbial community compositions in all the reactors after temperature disturbance was observed. More importantly, the three new steady-state microbial community compositions were clustered together and were clearly distinguished from the steady-state microbial community compositions before the temperature disturbances. A comparison of steady-state microbial community compositions before and after temperature disturbance showed that the temperature disturbance played an important role in the microbial community assembly and ecosystem function during the anaerobic digestion of cattle manure.

4.3 CONCLUSIONS

The present study showed that similar steady-state process performances and microbial community profiles were achieved in three biogas reactors with the same inoculum and operational conditions, which suggested a minor role of stochastic factors in shaping the profile of the microbial community composition and activity in biogas reactor. Instead, temperature disturbance played an important role in the microbial community composition as well as process performance in biogas reactors. Increased methane yields (around 10% higher) and decreased VFAs concentrations at steady states were found in all three reactors after the temperature disturbances, although three different temperature disturbances were applied to each biogas reactor. New steady-state microbial community profiles were also observed in all the biogas reactors after the temperature disturbance.

4.4 MATERIALS AND METHODS

4.4.1 INOCULUM AND SUBSTRATE

The inoculum (total solids (TS) 41.5 ± 1.4 g/L, VS 30.2 ± 1.2 g/L) used in this study was obtained from a mesophilic full-scale biogas plant (Hashoj biogas plant, Denmark) co-digesting cattle manure, pig manure, intestinal content from pig abattoirs, and fat and flotation sludge from pig abattoirs, fish and food processing industries, and so on. After collection, the inoculum was stored in an incubator with the temperature controlled at 37°C. The substrate used in the present study was cattle manure, and its characteristics were as follows: pH 7.5 ± 0.1, TS 36.6 ± 1.2 g/L, VS 25.2 ± 1.4 g/L, total nitrogen 2.2 ± 0.3 g/L.

4.4.2 REACTOR OPERATION

Three identical 1000-mL continuously stirred tank reactors (A, B, and C) with working volumes of 700 mL were used. The reactors were mixed by magnetic stirrer at a stirring speed of 150 rpm, and the produced biogas was collected by gas bags. All the reactors were put into an incubator with the temperature controlled at 37°C. The HRT of the reactors was 14 d. Initially, all the reactors were filled with the above-mentioned inoculum (600 mL) and cattle manure (100 mL). After one week's digestion, all the reactors were fed daily with cattle manure. Once steady states were achieved in all the reactors, the temperatures in the reactors were changed to 25°C, 45°C, and 55°C for reactors A, B, and C by putting them into different incubators with the corresponding temperatures, while the daily feeding with cattle manure continued. The temperatures were changed back to 37°C after 10 days' operation period by putting all the reactors back into the initial incubator with the temperature controlled at 37°C, and all the reactors were operated with daily feeding at 37°C, until new steady states were achieved. A steady state in this study was defined as a stable biogas production with a daily variation of lower than 10% for at least 10 days.

4.4.3 MICROBIAL ANALYSIS

Four samples were collected from the mixture in each reactor for microbial analysis. Samples A1, B1, and C1 were collected during the steady states of reactors A, B, and C before the temperature disturbance (day 50). Samples A2, B2, and C2 were collected 10 days after the temperatures were changed to 25°C, 45°C, and 55°C for reactors A, B, and C, respectively (day 60). Samples A3, B3, and C3 were collected 10 days after the temperatures of all the reactors were changed back to 37°C (day 70). Samples A4, B4, and C4 were collected during the steady states of reactors A, B, and C after the temperature disturbance (day 112).

The total genomic DNA of the collected samples was extracted using QIAamp DNA Stool Mini Kit (51504, QIAGEN, Valencia, CA) according to the manufacturer's instructions. The extracted DNA was amplified with the universal primers 515f and 806r [24],[25]. The forward primer was 5'-GTGCCAGCMGCCGCGGTAA-3', and the reverse primer was 5'-GGACTACHVGGGTWTCTAAT-3'. PCR conditions were set according to a previous study [24]. The PCR products were purified using the QIAquick spin columns (QIAGEN) to remove the excess primer dimers and dNTPs, and the concentration of PCR amplicons was measured by NanoDrop spectrophotometer [37]. The samples were then sent out for barcoded libraries preparation and sequencing on an Ion Torrent PGM system with 316 chip using the Ion Sequencing 400 bp Kit (all from Life Technologies) according to the standard protocol (Ion Xpress™ Plus gDNA and Amplicon Library Preparation, Life Technologies). The low-quality sequences without exact matches to the forward and reverse primers, with length shorter than 100 bp, and containing any ambiguous base calls, were removed from the raw sequencing data by RDP tools [38]. Chimeras were removed from the data by using the Find Chimeras web tool. The data were submitted to MG-RAST (ID 4556097.3-4556107.3). The sequences were phylogenetically assigned to taxonomic classifications by the RDP classifier with a confidence threshold of 50%. The PCoA and hierarchical cluster analysis were conducted by MG-RAST [39].

4.4.4 ANALYTICAL METHODS

TS, VS, and total nitrogen were analyzed according to the American Public Health Association (APHA) [40]. The concentrations of acetate, propionate, isobutyrate, butyrate, iso-valerate, and valerate were determined by a gas chromatograph (Hewlett Packard, HP5890 series II) equipped with a flame ionization detector and HP FFAP column (30 m × 0.53 mm × 1.0 μm). CH_4 was analyzed using a gas chromatograph with a thermal conductivity detector (GC-TCD) fitted with a parallel column of 1.1 m × 3/16 Molsieve 137 and 0.7 m × 1/4 Chromosorb 108. Detailed information about the operating conditions for the GC was given previously [41]. An analysis of variance was used to test the significance of results, and $P < 0.05$ was considered to be statistically significant. The methane yield was calculated based on the daily methane production (mL_{CH4}/d) divided by daily VS feeding (g_{VS}/d).

REFERENCES

1. Luo G, Wang W, Angelidaki I: Anaerobic digestion for simultaneous sewage sludge treatment and CO biomethanation: process performance and microbial ecology. Environ Sci Technol 2013, 47:10685-93.
2. Luo G, Angelidaki I: Co-digestion of manure and whey for in situ biogas upgrading by the addition of H2: process performance and microbial insights. Appl Microbiol Biotechnol 2013, 97:1373-81.
3. Werner JJ, Knights D, Garcia ML, Scalfone NB, Smith S, Yarasheski K, et al.: Bacterial community structures are unique and resilient in full-scale bioenergy systems. Proc Natl Acad Sci U S A 2011, 108:4158-63.
4. Chase JM: Stochastic community assembly causes higher biodiversity in more productive environments. Science 2010, 328:1388-91.
5. Chase JM: Drought mediates the importance of stochastic community assembly. Proc Natl Acad Sci U S A 2007, 104:17430-4.
6. Ofiteru ID, Lunn M, Curtis TP, Wells GF, Criddle CS, Francis CA, et al.: Combined niche and neutral effects in a microbial wastewater treatment community. Proc Natl Acad Sci U S A 2010, 107:15345-50.
7. Fargione J, Brown CS, Tilman D: Community assembly and invasion: an experimental test of neutral versus niche processes. Proc Natl Acad Sci U S A 2003, 100:8916-20.
8. Chave J: Neutral theory and community ecology. Ecol Lett 2004, 7:241-53.

9. Jiang L, Patel SN: Community assembly in the presence of disturbance: A microcosm experiment. Ecology 2008, 89:1931-40.
10. Trexler JC, Loftus WF, Perry S: Disturbance frequency and community structure in a twenty-five year intervention study. Oecologia 2005, 145:140-52.
11. Rincón B, Borja R, González JM, Portillo MC, Sáiz-Jiménez C: Influence of organic loading rate and hydraulic retention time on the performance, stability and microbial communities of one-stage anaerobic digestion of two-phase olive mill solid residue. Biochem Eng J 2008, 40:253-61.
12. Nielsen HB, Mladenovska Z, Westermann P, Ahring BK: Comparison of two-stage thermophilic (68°C/55°C) anaerobic digestion with one-stage thermophilic (55°C) digestion of cattle manure. Biotechnol Bioeng 2004, 86:291-300.
13. Chae KJ, Jang A, Yim SK, Kim IS: The effects of digestion temperature and temperature shock on the biogas yields from the mesophilic anaerobic digestion of swine manure. Bioresour Technol 2008, 99:1-6.
14. Boe K, Angelidaki I: Serial CSTR digester configuration for improving biogas production from manure. Water Res 2009, 43:166-72.
15. Nielsen HB, Ahring BK: Responses of the biogas process to pulses of oleate in reactors treating mixtures of cattle and pig manure. Biotechnol Bioeng 2006, 95:96-105.
16. Ahn JH, Forster CF: The effect of temperature variations on the performance of mesophilic and thermophilic anaerobic filters treating a simulated papermill wastewater. Process Biochem 2002, 37:589-94.
17. Ye L, Zhang T: Pathogenic bacteria in sewage treatment plants as revealed by 454 pyrosequencing. Environ Sci Technol 2011, 45:7173-9.
18. Loman NJ, Misra RV, Dallman TJ, Constantinidou C, Gharbia SE, Wain J, et al.: Performance comparison of benchtop high-throughput sequencing platforms. Nat Biotechnol 2012, 30:434-9.
19. Lescure JP, Delannoy B, Verrier D, Albagnac G. Consequence of a thermal accident on the microbial activity of an industrial anbaerobic filter. Proceedings of the 5th international conference on anaerobic digestion 1988:211–214.
20. Yu HQ, Fang HHP: Acidogenesis of gelatin-rich wastewater in an upflow anaerobic reactor: influence of pH and temperature. Water Res 2003, 37:55-66.
21. Demirel B, Scherer P: The roles of acetotrophic and hydrogenotrophic methanogens during anaerobic conversion of biomass to methane: a review. Rev Environ Sci Biotechnol 2008, 7:173-90.
22. Ma JX, Carballa M, Van de Caveye P, Verstraete W: Enhanced propionic acid degradation (EPAD) system: Proof of principle and feasibility. Water Res 2009, 43:3239-48.
23. Boe K, Batstone DJ, Steyer JP, Angelidaki I: State indicators for monitoring the anaerobic digestion process. Water Res 2010, 44:5973-80.
24. Bates ST, Berg-Lyons D, Caporaso JG, Walters WA, Knight R, Fierer N: Examining the global distribution of dominant archaeal populations in soil. ISME J 2011, 5:908-17.
25. Caporaso JG, Lauber CL, Walters WA, Berg-Lyons D, Lozupone CA, Turnbaugh PJ, et al.: Global patterns of 16S rRNA diversity at a depth of millions of sequences per sample. Proc Natl Acad Sci U S A 2011, 108:4516-22.

26. Sundberg C, Al-Soud WA, Larsson M, Alm E, Yekta SS, Svensson BH, et al.: 454 pyrosequencing analyses of bacterial and archaeal richness in 21 full-scale biogas digesters. FEMS Microbiol Ecol 2013, 85:612-26.
27. Riviere D, Desvignes V, Pelletier E, Chaussonnerie S, Guermazi S, Weissenbach J, et al.: Towards the definition of a core of microorganisms involved in anaerobic digestion of sludge. ISME J 2009, 3:700-14.
28. Lu L, Xing DF, Ren NQ: Pyrosequencing reveals highly diverse microbial communities in microbial electrolysis cells involved in enhanced H-2 production from waste activated sludge. Water Res 2012, 46:2425-34.
29. Debowski M, Korzeniewska E, Filipkowska Z, Zielinski M, Kwiatkowski R: Possibility of hydrogen production during cheese whey fermentation process by different strains of psychrophilic bacteria. Int J Hydrog Energy 2014, 39:1972-8.
30. Lay JJ: Biohydrogen generation by mesophilic anaerobic fermentation of microcrystalline cellulose. Biotechnol Bioeng 2001, 74:280-7.
31. Karakashev D, Batstone DJ, Angelidaki I: Influence of environmental conditions on methanogenic compositions in anaerobic biogas reactors. Appl Environ Microbiol 2005, 71:331-8.
32. Hao LP, Lu F, He PJ, Li L, Shao LM: Predominant contribution of syntrophic acetate oxidation to thermophilic methane formation at high acetate concentrations. Environ Sci Technol 2011, 45:508-13.
33. Fotidis IA, Karakashev D, Kotsopoulos TA, Martzopoulos GG, Angelidaki I: Effect of ammonium and acetate on methanogenic pathway and methanogenic community composition. FEMS Microbiol Ecol 2013, 83:38-48.
34. De Francisci D, Kougias PG, Treu L, Campanaro S, Angelidaki I. Microbial diversity and dynamicity of biogas reactors due to radical changes of feedstock composition. Bioresour Technol. 2015;176:56-64, doi: http://dx.doi.org/10.1016/j.biortech.2014.10.126.
35. Zhou JZ, Liu WZ, Deng Y, Jiang YH, Xue K, He ZL, et al. Stochastic Assembly Leads to Alternative Communities with Distinct Functions in a Bioreactor Microbial Community. Mbio 2013;4:e00584.
36. Pagaling E, Strathdee F, Spears BM, Cates ME, Allen RJ, Free A: Community history affects the predictability of microbial ecosystem development. ISME J 2014, 8:19-30.
37. Zhang H, Banaszak JE, Parameswaran P, Alder J, Krajmalnik-Brown R, Rittmann BE: Focused-Pulsed sludge pre-treatment increases the bacterial diversity and relative abundance of acetoclastic methanogens in a full-scale anaerobic digester. Water Res 2009, 43:4517-26.
38. Cole JR, Wang Q, Cardenas E, Fish J, Chai B, Farris RJ, et al.: The ribosomal database project: improved alignments and new tools for rRNA analysis. Nucleic Acids Res 2009, 37:D141-5.
39. Meyer F, Paarmann D, D'Souza M, Olson R, Glass EM, Kubal M, et al. The metagenomics RAST server - a public resource for the automatic phylogenetic and functional analysis of metagenomes. BMC Bioinformatics 2008;9:386.
40. APHA: Standard methods for the examination of water and wastewater, 19 th ed. American Public Health Association, New York, USA; 1995.

41. Luo G, Talebnia F, Karakashev D, Xie L, Zhou Q, Angelidaki I: Enhanced bioenergy recovery from rapeseed plant in a biorefinery concept. Bioresour Technol 2010, 102:1310-3.

There is one supplemental file that is not available in this version of the article. To view this additional information, please use the citation on the first page of this chapter.

PART II

COMPOSTING

CHAPTER 5

Composting of Organic Fraction of Municipal Solid Waste: A Pilot Plant in Maxixe District, Mozambique

C. COLLIVIGNARELLI, A. PERTEGHELLA, AND M. VACCARI

5.1 INTRODUCTION

Maxixe is the largest city and economic capital of the province of Inhambane, Mozambique, with a population of 121,097 inhabitants. Maxixe town is divided in urban, sub-urban and rural area for 20.3%, 35.6% and 44.1% respectively. The absence of proper waste management (Figure 1) in Maxixe District (Mozambique) is a common issue that has to be faced, like in many other African countries (Hoornweg and Bhada-Tata, 2012). Usually waste are abandoned along the streets in the center of the city or are often burned directly inside the municipal street container in order to empty them, reducing the waste volume. As stated by Cointreau (2006), lack of proper waste management, especially concerning the organic frac-

Collivignarelli C, Perteghella A, and M. Vacchari M. "Composting of Organic Fraction of Municipal Solid Waste: A Pilot Plant in Maxixe District, Mozambique." Venice 2014, Fifth International Symposium on Energy from Biomass and Waste, San Servolo, Venice, Italy; 17 - 20 November 2014. *CISA Publisher (2014). Used with permission from the publisher.*

tion of municipal solid waste (OFMSW), causes health risks for the population due to proliferation of disease carrying vectors (rats, mosquitoes, flies, birds, etc.), direct contact with waste (mainly for children and waste pickers), air pollution through indiscriminate burning, blocking of drains and flooding, water and soil pollution and greenhouse gas emissions.

According to Fernando and Carmo Lima (2012), the OFMSW of Maxixe district and Chambone, Rumbana and Malalane neighborhoods is approximately 50% of the total amount of produced waste, as showed in Table 1; it is coherent with the organic waste percentage of low, middle and upper income countries (Table 2). Even the percentage of paper, plastic, glass and metal of Maxixe district are comparable with data showed in Table 2.

TABLE 1: Composition of MSW in Maxixe district and in 3 others Maxixe neighborhoods (Fernando and Carmo Lima, 2012)

Monitored Location	Organic (%)	Paper (%)	Plastic (%)	Glass (%)	Metal (%)	Other (%)
Maxixe district	48	4	2	4	3	39
Chambone neighborhood	40	7	4	4	5	41
Rumbana neighborhood	58	3	2	4	2	32
Malalane neighborhood	46	2	1	3	3	45
Average	48	4	2	3	3	39

TABLE 2: Composition of MSW related to population income levels (Hoornweg and Bhada-Tata 2012)

Income Level	Organic (%)	Paper (%)	Plastic (%)	Glass (%)	Metal (%)	Other (%)
Low Income	64	5	8	3	3	17
Low Middle Income	59	9	12	3	2	15
Upper Middle income	54	14	11	5	3	13
High Income	28	31	11	7	6	17

FIGURE 1: Inadequate waste management practices in Maxixe District.

FIGURE 2: Pilot plant wood boxes.

As pointed out by Table 1, Maxixe district needs to treat the organic fraction properly, thus the composting process seems to be an appropriate solution in order to avoid environmental and sanitary problems. The composting process (Mansoor er al., 2004) allows transforming the organic matter into compost, valuable product that improves the soil quality (Gajalakshmi and Abbasi, 2008) from a physical-chemical point of view, enhancing at the same time the agronomic production and food security (Edmondson et al., 2014; Mdluli et al., 2013). The composting technology is widely used in developing countries, and it's possible to distinguish 3 main different scales of technology management: i) household composting process, ii) decentralized/community composting plants, iii) centralized composting plants. Each plant can treat the OFMSW with 2 different types of aerobic process, the windrow type (Rytz, 2001) and the vessel/bin/barrel type (Zurbrügg et al., 2004). Nevertheless scientific literature presents a big gap concerning data and information of composting plants running in low and middle income countries. This entails difficulties when decision makers want to implement the composting solution, with low possibilities to take as example other field experiences.

CeLIM, an Italian NGO, has implemented a project, which is still going on, with the aim to reduce the environmental and sanitary problems, caused by inadequate waste management, in Maxixe district. To reach the project aim, a lot of awareness campaigns were carried out for the population, and new street containers were introduced in order to increase the volume of waste collection and reduce the uncontrolled dumping along the streets. Moreover, a little centralized composting plant was built, near Maxixe dump in order to recover the organic fraction of municipal solid waste and produce compost to reduce the environmental burden, reduce the chemical fertilizer use and increase the soil quality towards sustainable agriculture and food security. CeTAmb LAB (Research laboratory on Appropriate Technologies for Environmental Management in resource-limited Countries, University of Brescia) has been involved for the study and monitoring of the pilot plant in order to assess its treatment capacity and increase its operational management.

This paper reports the results based on a comparison of two heaps realized at the composting pilot plant. The authors have compared and discussed the temperature trend and the physical-chemical analyses carried out on the final compost. The obtained results will be applied and test on the full scale plant, designed to treat 6 tons of organic waste per month.

5.2 MATERIALS AND METHODS

5.2.1 PILOT AREA

The composting pilot plant was built before the full scale plant, close to the Maxixe open dump, in order to represent approximately the same characteristics of the full scale plant, especially considering 2 working hypotheses: a duration equal to 100 days for the biological degradation process and an heaps overturning with frequency equal to 10 days. Ten 1 m3 wood boxes (Figure 2) were employed to guarantee the respect of the predetermined hypothesis. Different tests were performed on the pilot plant in order to understand what kind of organic waste mix is more suitable for the initial heap composition and its process, as well as to monitor the trend temperature and the amount of water required during the aerobic process.

5.2.2 EXPERIMENTAL ACTIVITY

Two heaps, named A and B, were considered to investigate how different heaps overturning frequencies influence the temperatures trend, the degradation process duration, and the quality of the final product. As shown in Table 3, the heaps were composed with the same organic substrate percentages and with the same initial mass.

TABLE 3: Initial amount of waste and waste composition of the analyzed heaps.

	Unit	Heap A	Heap B
Initial mass of the heap	kg	282.5	267.4
Garden pruning	%	49	47
Food waste	%	23	24
Coconut sawdust	%	17	18
Compost residue from sieving	%	11	11

Heap A and heap B were overturned every 10 and 5 days respectively. All the operations as heap formation, grinding, heap overturning, watering and sieving were manually performed. An operator collected initial heaps mass, heaps temperature, environmental temperature and heaps overturning and he has reported them on a monitoring form according to the date in which measurement and operations were performed.

The equipment used for the measurement was: i) 1 m long temperature probe (for heap temperature), ii) standard electronic thermometer (for environmental temperature), iii) mechanical balance (for initial and final heap mass). The heap temperature was measered putting the temperature probe inside the heap from the top layer, considering 3 measuring points: one in the heap core, one on the left and one on the right of the core point.

Due to difficulties in finding a chemical laboratory close to Maxixe district, the compost samples of heap A and B were analyzed in an Italian chemical laboratory, according to the Italian procedures and regulation laws (D. Lgs. 75/2010).

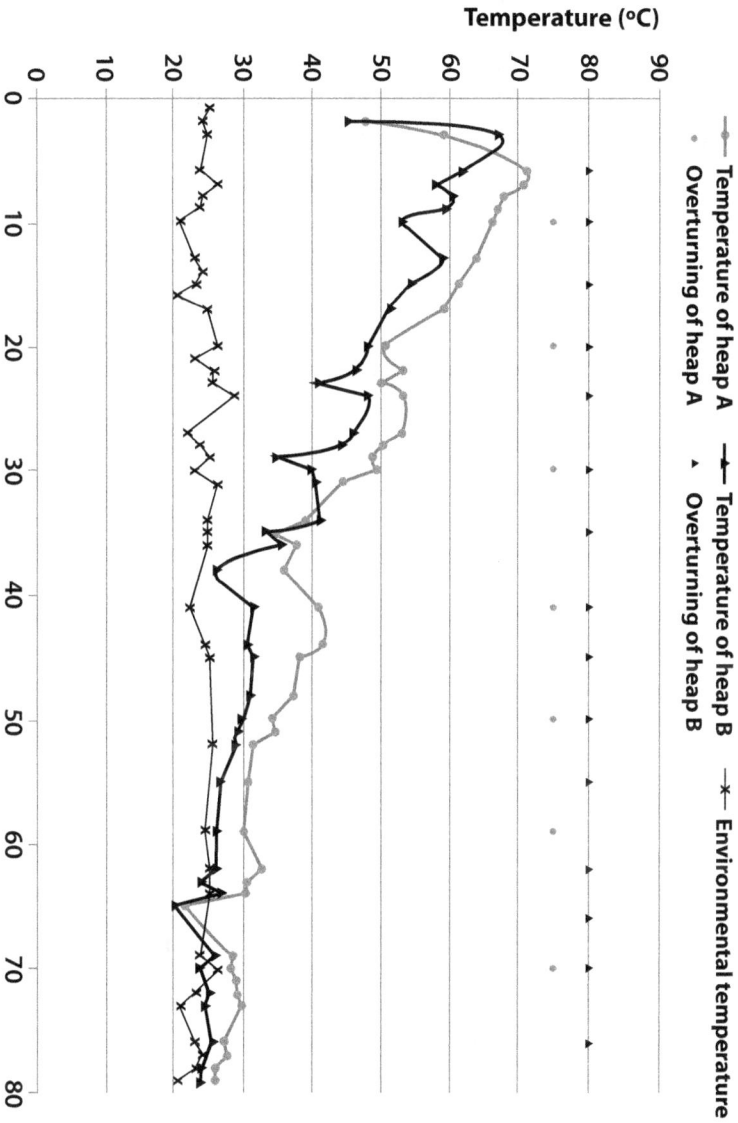

FIGURE 3: Temperature trend and frequency overturning: comparison between heap A and B.

5.3 RESULTS AND DISCUSSION

Heap A and B were removed from the wood box after 79 days of biological treatment. Subsequently the final product was spread out on the concrete basement in order to reduce the water content by natural drying. Then the compost was sieved and the final amount of saleable compost was 123.3 kg and 111.4 kg for heap A and B respectively. The compost production ratio, that are the kilograms of final compost produced per each kilograms of organic waste launched at the process, was 44% and 42% for heap A and B, respectively. Moreover, the water amount dosed during the process, necessary to control heaps temperature and moisture, was approximately 0.013 L/kg for both the heaps. Apparently the amount of water used seems to be adequate because no apparent problems raised during the process, as confirmed by the temperature trend (Figure 3).

Nevertheless it has not been possible to make a comparison with other scientific data because there are not similar data in scientific literature, and at the same time this parameter strongly depends on the process management and the climate.

5.3.1 TEMPERATURE TREND

Figure 3 shows the core temperature trend with time of the heap A and B, according with different overturning frequencies. Both of the temperature heaps increase rapidly in the first days, reaching the peak up to 70 °C within 1 week from the test launching. After reaching the peak, the temperature decreased gradually and constantly to the environmental temperature that stated out the end of the process after 80 days approximately.

According to Figure 3, heap A and B had the same temperature trend, but since for all the process duration the temperature trend of heap B was lower than the temperature trend of heap A, heap B reached the environmental temperature about a week before the heap A. A more frequent heap overturning (every 5 days) allows increasing the oxygen supply intensifying the biological degradation rate and reducing the duration process. Moreover it allows supervising the temperature, avoiding temperature

peaks higher than 82°C that could severely impede the microbial community (Gajalakshmi and Abbasi, 2008).

TABLE 4: Characteristics of the final compost produced in the two experimental runs.

Parameters	Units	Heap A	Heap B	Italian law quality parameters
Organic carbon	% DM	23.96	16.47*	> 20
Total nitrogen	% DM	0.71	0.42	-
Organic nitrogen	5% N tot	87.35	90.64	> 80% tot. N
Humic and fulvic acids	% DM	4.11*	2.44*	> 7
C/N	-	33.74*	39.21*	< 25
pH	-	8.6*	8.71*	6-8.5
Moisture	% t.q.	38.08	22.91	< 50
Dry matter	% t.q.	61.92	77.09	> 50
Cu	mg/kg DM	9.13	8.09	< 230
Zn	mg/kg DM	80.19	59.59	< 500
Cd	mg/kg DM	0.55	0.12	< 1.5
Ni	mg/kg DM	5.81	8.68	< 100
Hg	mg/kg DM	0.16	0.15	< 1.5
Cr^{IV}	mg/kg DM	0.33	Below Detection Limit	< 0.5
Salmonella	MPN/g	Absence	Absence	No presence
Escherichia Coli	CFU/g	Absence	Absence	< 1000
Germination Index	%	94.07	72.02	> 60

* Not in accordance with Italian law quality parameters set by Legislative Decree 2010/75

5.3.2 CHEMICAL ANALYSES

Table 4 point out the results of the physical-chemical analyses carried out
on two compost samples taken from the heaps A and B. All the results
comply with Italian law parameters, except the organic carbon and humic
and fulvic acids that are lower while carbon-nitrogen ratio (C/N) and pH
are higher than the prescribed value. Globally the chemical analyses show

good compost quality for both the heaps. All the measured parameters respect the Italian law limit, except for organic carbon, humic and fulvic acids, carbon-nitrogen ratio (C/N) and pH. Concerning the organic carbon measured in the heap B, the lower value was probably due to the low content of organic carbon in the initial organic waste, but at the same time due to the more frequent overturning that increased the organic carbon volatilization rate. Moreover the low content of organic nitrogen, for both the analyzed samples, determines a compost C/N ratio a little bit higher than the Italian limit. As concerns humic and fulvic acids, the low values measured in both of the heaps are probably due to the incomplete process of organic carbon conversion into humic and fulvic acids, that at the same time could depend on the organic carbon concentration in the initial used substrates. The low humic and fulvic acid concentrations contribute to have a basic pH of the produced compost. All the analyzed heavy metals (Cu, Zn, CD, Ni, Hg, CrVI) are below the Italian standard, and the good germination test results confirm the absence of uncontrolled heavy metal pollution.

Moreover the mean temperature reached by heap A and B in the first 20 days (between 50-70 °C) was fundamental for the pathogenic bacteria control (Gajalakshmi and Abbasi, 2008), indeed the microbiological analyses confirm the complete deactivation of *Salmonella* and *Escherichia coli* (Table 4), guaranteeing the microbiological safety of the use of that compost in agriculture.

5.4 CONCLUSIONS

The tests carried out on the composting pilot plant show how two different heap overturning frequencies can influence the composting process in term of duration, temperature and final product quality. As pointed out by heap B results, a more frequent heap overturning (every 5 days) allows to reduce the temperature trend, avoiding possible temperature peaks that could severely impede the microbial community, reduce the biological process duration (increasing the oxygen supply increase the biological degradation ratio) and therefore the treatment time of the material in the processing plant, allowing an increasing of the daily quantity of treated waste. At the same time an intensive heap overturning can cause an excessive volatilization of organic parameters as shown by heap B concerning the organic carbon and humic and fulvic acids that are below the Italian law limits. Anyway, even if the organic carbon (heap B), humic and fulvic acids (heap A and B), C/N ratio (heap A and B) and pH (heap A and B) do not respect the Italian standard limit, the quality of compost produced was good because there was not heavy metal pollution and at the same time the germination test shown good results.

The achieved results from the pilot tests represent a good starting point for the start-up of the full scale composting plant, nevertheless they need to be applied at the full scale plant in order to verify and correct them properly according to the composting process requirements for a good quality compost production.

REFERENCES

1. Cointreau S. (2006). Occupational and Environmental Health Issues of Solid Waste Management. Available at http://siteresources.worldbank.org/INTUSWM/Resources/up-2.pdf. The World Bank.
2. Edmondson J. L., Davies Z. G., Gaston K. J., Leake J. R. (2014).Urban cultivation in allotments maintains soil qualities adversely affected by conventional agriculture. J Appl Ecol, article in press.
3. D. Lgs. 29 aprile 2010, n.75. "Riordino e revisione della disciplina in materia di fertilizzanti, a norma dell'articolo 13 della legge 7 luglio 2009, n. 88". Pubblicato nella Gazzetta Ufficiale n. 121 del 26 maggio 2010.
4. Fernando A., Carmo Lima S. (2012). Caracterizacao dos residuos solidos urbanos do municipio de Maxixe/Mozambique. Caminhos de geografia. ISSN 1678-6343. Available at http://www.seer.ufu.br/index.php/caminhosdegeografia
5. Gajalakshmi S., Abbasi S. A. (2008). Solid Waste Management by Composting: State of the Art, Critical Reviews in Environmental Science and Technology. Available at http://dx.doi.org/10.1080/10643380701413633
6. Mansoor A., Harper M., Pervez A., Rouse J., Drescher S., Zurbrügg C., (2004). Sustainable Composting. Case studies and guidelines for developing countries. Available at http://r4d.dfid.gov.uk/pdf/outputs/R8063.pdf. WEDC, Loughborough University, England.
7. Mdluli F., Thamaga-Chitja J., Schmidt S. (2013). Appraisal of Hygiene Indicators and Farming Practices in the Production of Leafy Vegetables by Organic Small-Scale Farmers in uMbumbulu (Rural KwaZulu-Natal, South Africa). Int J Environ Res Public Health, vol. 10, 4323-4338.

8. Rytz I. (2001). Assessment of a decentralised composting scheme in Dhaka, Bangla-desh. Technical, operational, organizational and financial aspects. Available at http://www.eawag.ch/forschung/sandec/publikationen/swm/dl/Rytz_2001.pdf

9. Zurbrügg C., Drescher S., Patel A., Sharatchandra H. C. (2004). Decentralised com-posting of urban waste – an overview of community and private initiatives in Indian cities. Waste Manage vol. 24, 655–662.

10. Zurbrügg C., Gfrerer M., Ashadi H., Brenner W., Küper D. (2012). Determinants of sustainability in solid waste management – The Gianyar Waste Recovery Project in Indonesia. Waste Manage vol. 32, 2126-2133.

CHAPTER 6

Changes in Bacterial and Fungal Communities across Compost Recipes, Preparation Methods, and Composting Times

DEBORAH A. NEHER, THOMAS R. WEICHT, SCOTT T. BATES, JONATHAN W. LEFF, AND NOAH FIERER

6.1 INTRODUCTION

Municipalities, industry and agricultural farms are generating substantial amounts of organic wastes. These wastes not only strain landfill space, but also pose serious threats to the environment. Compostable materials (including paper, food wastes, and grass clippings) comprised 62% (155 million tons) of this waste stream in the US [1]. Composting represents an important solution for a more sustainable management of organic waste. Not only does composting remove waste, it can effectively convert the waste into a nutrient-rich organic amendment for a variety of agricultural, horticultural or landscaping applications. Countries, states, and municipalities are increasingly enacting legislation and regulation to

Changes in Bacterial and Fungal Communities across Compost Recipes, Preparation Methods, and Composting Times. © Neher DA, Weicht TR, Bates ST, Leff JW, and Fierer N. PLoS ONE 8,11 (2013). http://journals.plos.org/plosone/article?id=10.1371/journal.pone.0079512. Licensed under a Creative Commons Attribution 3.0 Unported License, http://creativecommons.org/licenses/by/3.0/.

promote the diversion of organics from solid waste disposal facilities to recycling and composting.

Composting is a controlled aerobic process that degrades organic waste to stable material, with the resident microbial community mediating the biodegradation and conversion processes. There are three distinct successional phases driving chemical and microbial changes through time, phases that are determined primarily by changes in temperature [2]: mesophilic phase (moderate temperatures rising to ~45°C), thermophilic phase (high temperatures peaking at ~70°C), curing phase (cooling to ambient temperature). Compost recipes can vary widely. For example, carbon sources can include straw, paper, woodchips, sawdust, or bark; whereas nitrogen sources can include animal manures, sewage sludge, and/or municipal solid waste. Large-scale commercial composting requires a high-temperature phase designed to facilitate the removal of human and plant pathogens. The primary types of commercial composting methods are windrow, aerated static pile, and vermicomposting. Although vermiculture does not inherently include a thermophilic phase [3], [4], material can be pre-composted through aerated static piles and windrow to remove substances toxic to earthworms, inactivate plant seeds and remove human and plant pathogens. Generally, we have a poor understanding of the biological dynamics that occur during the composting process and there are no current regulations or guidelines that define desirable microbiological properties of compost.

Microbial communities found in compost have not been well characterized and relatively few studies describe both their bacterial and fungal diversity even though both groups are likely important mediators of the composting process. Most studies have utilized culture-based methods [2], [4], [5], [6], [7] that are known to only capture a small portion of the microbial diversity found in environmental samples [8]. While a few studies have employed cultivation-independent approaches, a comprehensive perspective on compost microbial dynamics is still lacking because these studies have focused on only one of the three compost phases [9], [10], [11], [12], [13], [14] or have used fingerprinting techniques that offer limited taxonomic resolution [15], [16], [17], [18], [19], [20], [21]. The few available high-throughput sequencing studies have focused primarily on bacteria [22], [23], [24], [25] even though fungi likely play an important role in the compost process and may enhance the quality of compost. Overall, we still have a limited understanding of how microbial communities are influenced by different composting recipes or methods, or the phases of heating and cooling that occur during the compost process.

This study represents a comprehensive assessment of both bacteria and fungi associated with compost using high-throughput sequencing on the Illumina MiSeq platform. We investigated the influence of composting recipe and process on the structure of microbial communities, and how communities change through time when compost is produced on a commercial-scale.

6.2 METHODS

6.2.1 FIELD SITE AND SAMPLING

To insure validity and applicability of results, all compost recipes were produced at a commercial compost production facility at Highfields Center for Composting (HCC; Hardwick, Vermont). All recipes contained 25% manure-silage from the same source, and all composting processes included a thermophilic phase. For all experiments, a sample is defined as a composite of 10 subsamples collected at random depths from a given pile that are mixed to be representative of a pile. Samples were frozen at −20°C until processed.

6.2.2 EXPERIMENT 1: COMPOST RECIPES

Three recipes contained manure from the same source with varying carbon sources were prepared and composted using the aerated static pile process. Recipe one was 100% manure-silage from a typical Vermont dairy barn, with a C:N ratio of 17:1, as a control. Recipe two used hay as a carbon source, and was mixed in a 3:1 ratio (volume basis) with manure/silage resulting in a C:N ratio of 23:1. Recipe three used hardwood as a carbon source mixed in a 5:5:3 ratio of manure/silage:hardwood bark: softwood shavings resulting in a C:N ratio of 34:1. During the early curing phase, compost was delivered to two farm locations where it continued to cure for three months prior to sampling. In total, eight samples were taken from each of the three recipes for a total of 24 samples. Hereafter, the recipes will be referred to as 'manure-silage', 'hay', and 'hardwood', respectively.

6.2.3 EXPERIMENT 2: COMPOSTING PROCESS

A standard commercial recipe was used in each of the three composting processes: windrow, aerated static pile, vermicompost. This recipe was comprised of 20% food residuals, 10–15% 2.5 cm woody material (e.g., hardwood bark and mixed wood chips), 10% hay, up to 5% shredded paper, up to 2% dry sawdust or shavings, and 50–60% mixed livestock manures (e.g., horse, cow, heifer, calf) mixed with various bedding materials (e.g., straw and hay). Four replicate samples were taken per treatment, for a total of 12 samples. Total genomic DNA was extracted from two sub-samples for each of two piles per composting process. Windrow, aerated static pile, and vermicompost samples were collected after curing, i.e., 9, 6, and 7 months of composting, respectively.

6.2.3.1 WINDROW

Windrow involves placing a mixture of organic waste materials into long, narrow piles on a composting pad which are turned frequently [3]. Piles were mixed with a bucket loader and were capped with manure/bedding to meet Vermont and National Organic Standards (NOS) Board regulations (www.ams.usda.gov/nop) and the piles were managed to maintain a temperature between 55–77°C for a minimum of 15 days and turned with a bucket loader a minimum of five times to ensure all materials have been subjected to the minimum temperature requirements.

6.2.3.2 AERATED STATIC PILE (ASP)

ASP systems force air throughout the pile and does not require turning once the pile has been formed, thus allowing for larger piles to be produced [3]. Piles were mixed with a bucket loader, placed in a three-sided ASP bay, and capped with manure/bedding. Each ASP pile was built to a height of 2 to 2.5 m after settling. Initial piles were 57–76 m^3 when placed in ASP bays. Piles were aerated in place with an in-floor air delivery system that uses a 20 cm layer of wood chips between the duct-work and pile to evenly distribute air. Blower fans were managed with speed control and timers, to meet 'Process to Further Reduce Pathogens' (PFRP) requirements as dictated by Vermont and NOS regulations. Briefly, piles were aerated in the ASP system for 3–6 weeks so the pile attained temperatures of 55°C for a minimum of three days. Piles were then re-stacked on composting pads, and turned regularly to continue composting.

6.2.3.3 VERMICOMPOST

Vermicompost is a mesophilic process that employs earthworms to stabilize organic residues [4]. Material entering the Continuous-Flow Worm Reactor was taken from piles on the composting pad, after they have been through the ASP procedure (outlined above), and re-stacked on the pads. Material had already met PFRP, and was generally four to six weeks old. Fresh material (0.76 m^3) was spread out weekly in a 3.8 cm layer on top of the bed, where it was allowed to continue to decompose, and be consumed by earthworms (*Eisenia fetida*). The bed was 1.52 m wide, 12.19 m long, and 0.6 m deep. The worm bed was housed in an indoor, heated room, and the compost temperature was 21–27°C. The compost remained in the worm bin for 60 to 90 days, the time it took for the fresh compost to be decomposed and move downward and out of the bottom of the bed.

6.2.4 EXPERIMENT 3: CHANGES IN MICROBIAL COMMUNITIES DURING THE COMPOSTING PROCESS

A standard commercial recipe of HCC was composted through windrow piles, aerated static pile, and vermicompost. Samples were collected on 20 September 2012 and represent various ages throughout the thermophilic and curing phases of the compost process for windrow, aerated static pile, and vermicomposting. Duplicate samples were analyzed at each time point for each composting method yielding a total of 24 samples.

6.2.5 DNA EXTRACTION, PCR AMPLIFICATION, SEQUENCING

Genomic DNA was extracted using the MoBio PowerSoil™ kit (MoBio, Carlsbad, CA, USA) according to the manufacturer's instructions following the method described in Lauber et al. [26]. PCR amplification of the 16S rRNA gene (for bacteria and archaea) or the internal transcribed spacer region (ITS1) of the nuclear ribosomal RNA gene (for fungi) followed the approach described in Fierer et al. [27]. Briefly, each sample was amplified in triplicate, and amplicons were composited together in equimolar concentrations prior to sequencing. PCR reactions contained 13 μL PCR-grade water, 10 μL 5 Prime Hot Master Mix, 0.5 μL each of the forward and reverse primers (10 μM final concentration), and 1.0 μL genomic DNA (diluted 1:10 with PCR-grade water). Reactions were held at 94°C for 3 min to denature the DNA, with amplification proceeding for 35 cycles at 94°C for 45 s, 50°C for 60 s, and 72°C for 90 s; a final extension of 10 min at 72°C was added to ensure complete amplification. For the bacterial and archaeal analyses, the PCR primers (515f/806r) targeted the V4 region of the 16S rRNA gene [28]. For the fungal analyses, we used PCR primers (ITS1-F/ITS2) to amplify the ITS1 spacer [29]. Both primer pairs contained 12-bp barcodes unique to each sample and the appropriate adapters to permit sequencing on the Illumina MiSeq platform [28], [30].

6.2.6 DATA ANALYSIS

Quality filtering, assignment of sequences to samples based on their barcodes, and clustering of sequences into operational taxonomic units (OTUs) was done following the standard QIIME pipeline [31] with sequence data quality-filtered as described previously. OTUs were determined using an open reference-based approach that implements reference-based clustering followed by de novo clustering using the UCLUST algorithm [32]. Clustering was conducted at the 97% similarity level using pre-clustered versions of the October 2012 Greengenes database (for 16S rRNA) [33] and November 2012 UNITE database (fungal ITS gene) [34] for the sequence reference set. Fungal ITS sequence processing followed the procedure outlined in McGuire et al. [30]. Sequences were assigned to taxonomic groups using the RDP classifier [35]. To keep sequencing depth consistent across all samples, the sequence data were rarified by randomly subsampling 2,000 and 100 reads per sample before downstream analyses of the 16S and fungal ITS rRNA datasets, respectively. Amplicon sequences were deposited in the public EMBL-EBI database (http://www.ebi.ac.uk/) and may be accessed using the accession numbers, 'ERP003625' and 'ERP003626' for the 16S and ITS sequences, respectively.

6.2.7 STATISTICAL ANALYSIS

We used the PRIMER v.6 software package (PRIMER-E, Plymouth, WA, USA) [36] for the calculation of pair-wise differences in community composition (Bray-Curtis distances) and the subsequent analyses of the pair-wise dissimilarity matrices via principal coordinate analysis and permutational multivariate analysis of variance (PERMANOVA). We used PERMANOVA to assess the effects of compost type and process type on the composition of the bacterial and fungal community compositions. For compost type, we included the farm identity as a random factor in our model to account for variation between farms. Differences in the relative abundance of specific taxa across recipe and methods were determined using multiple Kruskal-Wallis tests in R [37] and applying false discovery rate corrections to p-values to account for the multiple comparisons. Tests were only performed for the more abundant taxa (those with median relative abundances greater than 1.0% in any of the recipes or processes).

6.3 RESULTS

We obtained a total of 799,030 and 35,280 150-bp quality-filtered 16S and fungal ITS rRNA gene sequences across all samples, respectively. The number of 16S rRNA sequences obtained per sample varied from 770 to

13,390 (median = 5,398), and the number of fungal ITS sequences per sample varied from 51 to 663 (median = 215).

TABLE 1: Mean ± 1 SD (n = 8) of dominant bacterial phyla and sub-phyla, expressed as percentage of sequences in cured manure, hay and hardwood compost recipes.

Taxon	Manure	Hay	Hardwood
*Acidobacteria***	1.2 ± 0.7	0.7 ± 0.1	7.4 ± 2.7
Actinobacteria*	4.9 ± 2.0	9.9 ± 3.8	6.8 ± 2.0
Bacteroidetes*	27.0 ± 4.8	27.7 ± 2.7	21.0 ± 3.6
*Chloroflexi***	5.5 ± 3.9	2.3 ± 0.6	11.8 ± 8.5
Firmicutes**	5.2 ± 2.7	6.8 ± 1.6	1.5 ± 0.5
Gemmatimonadetes**	1.9 ± 0.8	4.4 ± 0.6	2.9 ± 0.8
Planctomycetes[n.s.]	2.6 ± 1.4	1.8 ± 0.3	3.0 ± 0.7
α-Proteobacteria[n.s]	7.8 ± 1.6	6.2 ± 2.4	6.3 ± 1.7
β-Proteobacteria[n.s]	7.0 ± 3.7	4.1 ± 0.7	4.0 ± 1.9
δ-Proteobacteria[n.s]	7.0 ± 1.7	7.0 ± 2.3	7.2 ± 1.5
γ-Proteobacteria*	12.5 ± 5.9	13.5 ± 1.5	9.2 ± 2.8
Verrucomicrobia***	1.7 ± 1.0	1.5 ± 0.3	4.6 ± 1.1

False Discovery Rate (FDR) p-value from Kruskal-Wallis Test. [n.s]*:* $p_{FDR} > 0.05$. **: 0.01 <* $p_{FDR} < 0.05$. ***: 0.01 <* $p_{FDR} < 0.05$. ****: 0.001 <* $p_{FDR} < 0.01$.

6.3.1 COMPOST RECIPES

All pair-wise comparisons of recipes contained unique communities of bacteria ($p_{(perm)} = 0.001$)and fungi ($p_{(perm)} = 0.001$). There was far more variability in bacterial (Table S1) and fungal community composition between compost recipes than across replicate samples collected from the same recipe (Figure 1).

Proteobacteria and *Bacteroidetes* were the most abundant bacterial phyla among compost recipes (Table 1). Within *Proteobacteria*, the γ-proteobacteria was more abundant than α-proteobacteria, β-proteobacteria, and δ-proteobacteria. *Archaea* were rare (median 1.2%

of reads for all samples), comprised mostly of *Crenarchaeota* and *Euryarchaeota*. Similarly, manure and hay contained greater relative abundances of *Firmicutes* than hardwood. Uniquely, hay contained more *Actinobacteria* and *Gemmatimonadetes*. In contrast, hardwood contained a greater relative abundance of *Acidobacteria* and *Chloroflexi*.

Ascomycota was the most abundant fungal phylum among recipes, about nearly two-fold more abundant than the *Basidiomycota* (Table 2). *Basidiomycota* were represented by *Agaricomycetes* and two undefined taxonomic classes. Manure-silage was distinguished by containing the greatest abundance of *Pezizomycetes* (including *Ascobolus*) and *Microascales*. Uniquely, hay contained greater abundances of *Epicoccum, Thermomyces, Eurotium, Arthrobotrys*, and *Myriococcum* (Table 2). Manure-silage and hay, but not hardwood, contained *Chytridomycota* and *Zygomycota* with hardwood containing greater abundances of *Sordariomycetes*.

6.3.2 COMPOST PROCESS

Holding recipe constant, bacterial (p(perm) = 0.002, Fig. 2A) and fungal (p(perm) = 0.003, Figure 2B) communities varied according to the type of managed compost process (Table S2). The compost prepared by ASP and windrow harbored bacterial and fungal communities that were more similar to one another than to the vermicompost-treated compost (Figure 2).

Bacterial communities in the windrow method were characterized by having a greater relative abundance of *Chloroflexi* and *Chlorobi* (Table 3). In contrast, the product of the vermicompost process contained relatively abundant *Bacteroidetes, γ-Proteobacteria*, and *Verrucomicrobia*. Fungal communities of windrow were dominated by *Sordariomycetes, Acremonium,* and an unclassified group in the *Basidiomycota* (Table 4). ASP was distinguished by a greater relative abundance of an unclassified family of *Pezizales*. Vermicompost contained the greatest relative abundance of *Arthrobotrys, Microascaceae, Zopfiella, Agaricomycetes,* and *Mortierella*.

TABLE 2: Mean ± 1 SD of fungal genera, expressed as percentage of sequences classified to phylum level in cured manure, hay, and hardwood compost recipes.

Phylum	Class	Order	Family	Genus	Manure (n=8)	Hay (n=7)	Hardwood (n=8)
Ascomycota[ns]					61.5 ± 16.4	55.9 ± 14.7	69.4 ± 22.6
	Dothideomycetes*	Pleosporales*	Pleosporaceae**	Epicoccum**	0 ± 0	2.1 ± 1.5	0.3 ± 0.8
	Eurotiomycetes**	Eurotiales**	Unknown[b,ns]	Thermomyces[ns]	0 ± 0	2.7 ± 3.7	0 ± 0
		Eurotiales**	Trichocomaceae**	Eurotium**	0 ± 0	2.0 ± 2.0	0 ± 0
	Orbiliomycetes*	Orbiliales*	Orbiliaceae*	Arthrobotrys[*]	0.4 ± 0.9	6.1 ± 7.0	0 ± 0
	Pezizomycetes*	Pezizales*	Ascobolaceae*	Ascobolus*	9.3 ± 12.3	2.4 ± 3.7	0 ± 0
		Unknown*	Unknown*	Unknown*	3.2 ± 3.1	1.1 ± 1.3	0 ± 0
	Sordariomycetes*	Hypocreales*	Unknown**	Acremonium*	0 ± 0	1.4 ± 1.1	0 ± 0
		Microascales**	Microascacae**	Pseudallescheria[ns]	2.7 ± 3.7	0.2 ± 0.4	0 ± 0
		Microascales**	Microascacae**	Scedosporium[*]	0.2 ± 0.5	7.7 ± 16.1	0 ± 0
		Microascales**	Microascacae**	Unknown**	14.1 ± 12.9	1.8 ± 2.5	0 ± 0
		Microascales**	Unknown**	Unknown**	7.7 ± 7.4	0.5 ± 1.2	0 ± 0
		Unknown**	Unknown**	Unknown**	14.9 ± 10.6	7.0 ± 4.5	54.2 ± 16.7
		Sordariales*	Lasiosphaeriaceae[ns]	Zopfiella[ns]	0.3 ± 0.9	1.5 ± 2.8	3.4 ± 5.4
		Sordariales*	Lasiosphaeriaceae[ns]	Unknown**	0 ± 0	0 ± 0	2.3 ± 1.8
		Sordariales*	Unknown**	Unknown**	0 ± 0	0.2 ± 0.4	2.8 ± 2.6
	Unknown[ns]	Unknown[ns]	Unknown[ns]	Unknown[ns]	4.9 ± 3.4	5.3 ± 2.6	5.4 ± 4.4
Basidiomycota[ns]	Agaricomycetes[ns]	Agaricales[ns]	Psathyrellaceae[ns]	Coprinellus[ns]	29.0 ± 15.8	35.3 ± 16.7	30.3 ± 22.9
					0.3 ± 0.9	0 ± 0	6.6 ± 9.6

TABLE 2: *Cont.*

Phylum	Class	Order	Family	Genus	Manure (n=8)	Hay (n=7)	Hardwood (n=8)
		Agaricales[ns]	Psathyrellaceae[ns]	Coprinus[ns]	3.2 ± 6.9	0 ± 0	1.0 ± 2.1
		Agaricales[ns]	Psathyrellaceae[ns]	Unknown*	3.4 ± 3.9	1.5 ± 2.1	0 ± 0
		Corticiales[ns]	Corticiaceae[ns]	Unknown[ns]	0 ± 0	1.3 ± 1.4	2.5 ± 3.7
		Unknown**	Unknown**	Myriococcum**[e]	0 ± 0	9.8 ± 8.7	0 ± 0
		Unknown[ns]	Unknown[ns]	Unknown[ns]	0.7 ± 1.0	18.0 ± 15.5	18.4 ± 24.6
		Unknown**	Unknown**	Unknown**	16.8 ± 11.0	0 ± 0	0 ± 0
		Unknown*	Unknown*	Unknown*	3.2 ± 1.6	2.8 ± 2.8	0.8 ± 1.3
Chytridiomycota*					**6.7 ± 6.7**	**5.0 ± 5.6**	**0 ± 0**
	Chytridiomycetes*	Spizellomycetales[ns]	Spizellomycetaceae[ns]	Gaertneriomyces[ns]	0.3 ± 0.8	1.6 ± 4.2	0 ± 0
		Unknown*	Unknown*	Unknown*	4.8 ± 6.6	3.4 ± 3.3	0 ± 0
Zygomycota*					**2.9 ± 2.2**	**3.8 ± 3.2**	**0.3 ± 0.8**
	Incertae_sedis*	Harpellales*	Legeriomycetaceae*	Smittium*[f]	1.3 ± 1.5	0.3 ± 0.6	0 ± 0
		Mortierellales[ns]	Mortierellaceae[ns]	Mortierella[ns]	1.1 ± 1.6	1.1 ± 2.9	0 ± 0
		Mucorales[ns]	Mucoraceae[ns]	Mucor[ns]	0 ± 0	2.1 ± 2.9	0 ± 0

*False Discovery Rate (FDR) p-values from Kruskal-Wallis test. [ns].: $P_{FDR} > 0.05$, *: $0.01 < p_{FDR} < 0.05$, **: $0.001 < p_{FDR} < 0.01$. [a]: rank order of Epicoccum species abundance: E. sp_CHTAM7, E. sp_TMS_2011. [b] represents a) a sequence from an undescribed taxon, b) from an environmental sequence were the organism was not identified, or c) a sequence matches a described species that is not represented in the reference database. [c]: rank order of Arthrobotrys species abundance: A. amerospora>A. oligospora>A. flagrans >A. oligospora. [d]: rank order of Scedosporium species abundance: S. prolificans>S. apiospermum. [e]: dominant species: Myriococcum thermophilum. [f]: rank order of Smittium species abundance: Smittium sp.> S. orthocladii.*

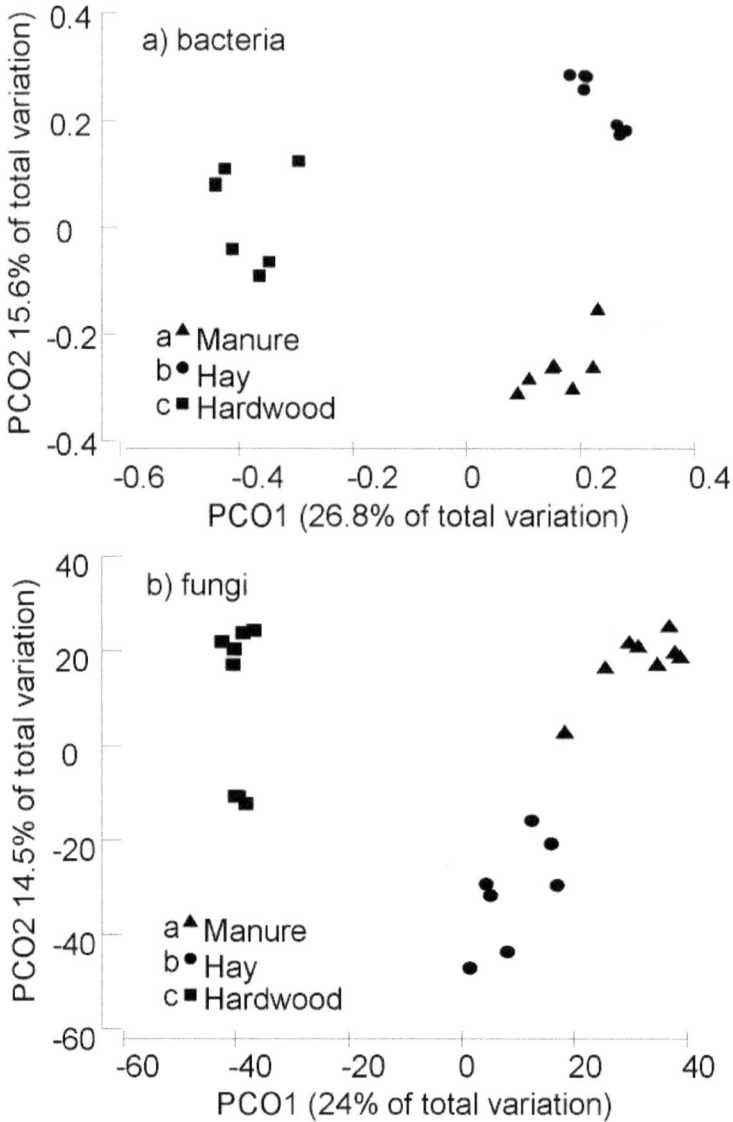

FIGURE 1: Principal coordinates analysis biplot for a) bacterial and b) fungal communities in three compost recipes (triangle: manure-silage, circle: hay, square: hardwood), n = 7 per treatment. Permutational multivariate analysis of variance indicated that differences between communities were highly significant (p≤0.001). Contrasting superscripts indicate that treatments are significantly different (p≤0.05).

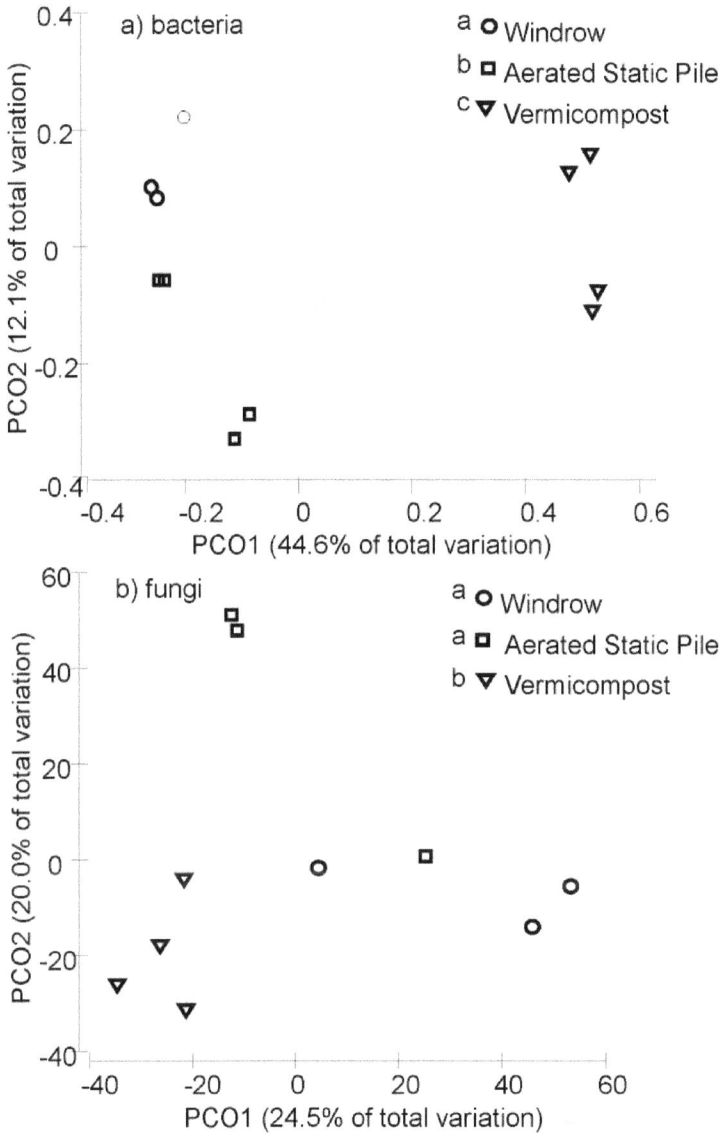

FIGURE 2: Principal coordinates analysis biplot for a) bacterial and b) fungal communities in the end product of three compost processes (circle: windrow, square: aerated static pile, inverted triangle: vermicompost), n = 4 per process. Permutational multivariate analysis of variance indicated that differences between communities were highly significant (p≤0.001). Contrasting superscripts indicate that treatments are significantly different (p≤0.05).

a) Bacteria

Taxa / Time	Windrow					Aerated Static Pile					Vermicompost		
	n=2	n=2	n=2	n=2	n=2	n=2	n=2	n=0	n=2	n=2	n=2	n=2	n=2
	155	191	239	281	323	22	56	155	160	183	53	143	203
Bacteroidetes	24.4	25.4	14.7	20.6	22.3	2.0	12.9		26.6	23.6	10.1	29.7	22.2
Chloroflexi	16.6	21.6	21.1	23.1	16.9	2.3	1.7		10.6	17.6	1.4	1.4	3.4
γ-Proteobacteria	11.5	10.4	11.2	8.7	9.9	4.6	27.8		15.9	9.9	44.3	17.2	15.9
δ-proteobacteria	10.7	7.0	8.3	7.6	6.2	2.5	3.6		9.6	7.7	1.6	11.5	7.2
Acidobacteria	2.0	3.0	5.5	5.4	4.9	0.1	0.5		1.9	4.3	0.2	1.7	4.3
Planctomycetes	2.3	2.7	3.9	2.7	4.8	0.5	0.9		2.7	3.5	0.3	3.2	4.2
β-Proteobacteria	4.5	4.1	4.5	3.2	4.6	3.7	3.5		6.4	3.0	5.7	6.2	8.2
α-Proteobacteria	2.3	2.9	3.9	4.2	4.6	2.1	6.6		3.5	3.7	3.9	3.8	5.4
Firmicutes	3.0	3.1	2.6	2.1	2.5	62.1	22.6		2.3	2.4	22.8	1.1	2.1
Actinobacteria	3.8	2.1	1.4	1.4	2.2	7.5	3.1		1.9	3.2	1.3	2.0	2.6
Verrucomicrobia	3.0	1.5	2.6	1.2	1.7	0.2	0.6		2.8	2.2	0.3	5.0	3.5
Gemmatimonadetes	2.6	3.1	4.2	2.1	1.6	3.9	2.5		1.7	2.6	0.5	1.2	2.1

b) Fungi (Ascomycota)

class	order	family	genus	Windrow					Aerated Static Pile					Vermicompost		
				n=2	n=0	n=2	n=1	n=2	n=1	n=2	n=2	n=2	n=2	n=0	n=2	n=2
				155	191	239	261	323	22	56	155	169	183	53	143	203
Dothideomycetes	Capnodiales	Mycosphaerellaceae[b]	Cladosporium	9.3		45.1	39.3	58.2	1.0	2.3	10.5	3.3	50.0	7.0	0.0	3.3
	Incertae_sedis[b]	Eremomycetaceae	Arthrographis	0.0		0.0	0.0	0.0	2.0	1.1	55.7	54.3	0.0	0.0	22.2	1.7
Eurotiomycetes	Eurotiales	Trichocomaceae	Thermomyces	10.9		1.2	9.0	1.3	0.0	5.0	12.6	14.6	0.0	6.6	0.0	8.7
	Eurotiales	Trichocomaceae	Penicillium	13.2		8.7	15.8	3.8	5.0	3.8	10.2	13.9	4.7	6.6	6.6	3.8
	Eurotiales	Trichocomaceae	Talaromyces	4.2		0.0	0.0	5.1	72.0	0.8	0.0	1.3	0.0	0.9	0.9	0.0
	Eurotiales	unknown	unknown	0.6		1.8	3.3	3.1	3.0	25.2	0.0	1.3	6.3	0.9	0.9	11.2
	unknown	unknown	unknown	4.2		33.3	0.0	0.0	0.0	0.8	0.0	0.0	0.0	0.0	16.8	6.1
Lecanoromycetes	unknown	unknown	unknown	0.0		0.6	0.0	0.0	0.0	0.0	0.0	2.0	0.0	0.0	0.0	14.9
Leotiomycetes	Helotiales	Incertae sedis	Scytalidium	18.2		0.0	2.7	1.3	0.0	2.4	2.3	0.7	0.0	0.0	0.0	0.6
	unknown	unknown	unknown	0.0		0.0	0.0	1.3	1.0	3.0	0.0	0.0	0.0	0.0	0.0	19.7
Orbiliomycetes	Orbiliales	Orbiliaceae	Arthrobotrys	2.9		0.0	2.6	1.3	0.0	1.6	3.4	2.6	4.7	3.0	3.0	1.8
Pezizomycetes	Pezizales	Ascobolaceae	Ascobolus	0.0		0.0	0.0	0.0	1.0	10.6	3.4	0.7	0.0	0.0	0.0	1.1
	Pezizales	Pezizaceae	unknown	0.0		0.0	0.0	0.0	0.0	1.5	0.0	0.0	7.8	16.4	16.4	0.0
	Pezizales	unknown	unknown	3.2		0.6	15.6	2.5	2.0	4.9	3.4	0.0	6.3	0.0	0.0	2.6
	unknown	unknown	unknown	0.0		0.0	0.0	2.5	0.0	0.0	0.0	0.0	0.0	0.0	0.0	0.0
Saccharomycetes	Saccharomycetales	Dipodascaceae	Galactomyces	0.0		0.0	0.0	0.0						12.6		1.9
Sordariomycetes	Hypocreales	Incertae_sedis[c]	Acremonium[c]	0.6		0.0	1.6	16.5		5.7				0.0	0.0	0.0
	Hypocreales	unknown	unknown	0.0		1.2	0.8	0.0		8.5				3.0		2.6
	Microascales	Microascaceae	Graphium	0.0		1.3	0.0	0.0						0.0		0.0
	Microascales	Microascaceae	Pseudallescheria	0.0		0.0	0.0	1.3	1.6				6.3	1.0		0.7
	unknown	unknown	unknown	0.6		0.0	0.0	0.0	5.7		0.8	0.7		1.0		1.3

[a] represents a) a sequence from an described taxon b) from an environmental sequence were the organism was not identified or c) a sequence matches a described species that is not represented in the reference database

[b] from a known genus, family, but the phylogeny is not resolved

[c] rank order of Acremonium abundance A_sp_ATT126 > A_alcalophilum > A_chrysogenum > A_rutilum

FIGURE 3: Heat map illustrating changes in A) bacterial and B) fungal composition through time for the same recipe composted by three processes: windrow, aerated static pile or vermicompost. All fungi illustrated are *ascomycota*. Time is expressed as days of decomposition. The thermophillic phase occurred prior to sampling in windrow, days 22–56 for aerated static pile, and day 53 for vermicompost. Units illustrated as mean percentages of total sequences (bacteria) and percentage of taxa classified to phylum (fungi). Dots represent missing samples.

TABLE 3: Mean ± 1 SD (n = 4) of total sequences classified as bacteria in a common recipe processed by windrow, aerated static pile or vermicompost.

	Windrow	Aerated Static Pile	Vermicompost
Bacteroidetes*	21.5 ± 4.6	16.3 ± 0.7	29.4 ± 10.5
Chlorobi*	2.9 ± 1.1	1.0 ± 0.2	0.2 ± 1.1
Chloroflexi*	19.8 ± 3.8	8.0 ± 1.2	2.4 ± 7.3
γ-Proteobacteria^	10.3 ± 1.3	14.6 ± 1.2	16.5 ± 9.7
Verrucombicrobia^	2.0 ± 0.8	1.5 ± 1.3	4.2 ± 1.2

Values are expressed as percentages. *: $p \leq 0.05$ false discovery rate (adjusted) from KW and unadjusted P-values. ^: $p \leq 0.05$ for unadjusted P-value, but ≤ 0.1 for false discovery rate (adjusted).

6.3.3 DYNAMICS OF COMPOSTING PROCESS

The temporal shifts in bacterial and fungal communities during the composting process were influenced primarily by whether the curing phase was managed as windrow, ASP or vermicompost (Figure 3). For the bacteria, *Bacteroidetes* varied by compost process through time (Figure 3A). *Bacteroidetes* were abundant following the thermophilic phase, but declined and subsequently increased in relative abundance as composting progressed in the windrow process but steadily increased through time in the ASP and vermicompost processes. The thermophilic phase of ASP and vermicompost was dominated by *Firmicutes*. The relative abundance of γ-*Proteobacteria* increased soon after the thermophilic phase and declined through time for all composting processes but to different extents. *Chloroflexi* were relatively abundant at the end of the thermophilic phase of ASP, fluctuated in abundance through time in windrow, and were uncommon in vermicompost.

Fungal community composition varied through time and in different trajectories for the three composting processes (Figure 3B). In windrow, *Thermomyces* and *Scytalidium* were the dominant fungi after the thermophilic phase and declined through time whereas *Cladosporium* increased through time (Table 3b). *Acremonium* was second to *Cladosporium* in dominance within the finished compost. *Talaromyces* dominated the thermophilic phase of ASP while *Arthrographis* and *Cladosporium* predominate later in the ASP process. Vermicompost had approximately equal abundances of T*hermomyces, Pezizaceae, Galactomyces,* and *Lecanoromycetes* with similar abundances of *Lecanoromycetes* remaining in finished compost, accompanied by *Eurotiales* and *Leotimycetes*.

6.4 DISCUSSION

As the most comprehensive assessment of compost bacterial and fungal communities conducted to date, this work provides unique insight into microbial dynamics across different compost recipes, preparation techniques, and through time as compost cures. The types of bacterial and fungal taxa found in compost with the high-throughput sequencing methods employed here are similar to previous studies that have used other approaches to describe microbial communities in composts of similar feedstock and/or process as in this study [2], [7], [10], [12], [23], [25], [38]. For example, we found large numbers of sequences for *Bacteroidetes, Proteobacteria, Acremonium, Ascobolus,* and *Mortierella,* taxa that have been commonly associated with compost. Furthermore, *Aspergillus, Penicillium, Mucor,* and *Alternaria* were present as common saprophytic fungi on food wastes [2], [4], [5].

6.4.1 COMPOST RECIPE

There are distinct types of microbial communities in finished compost products that originate from different source materials. We found taxa similar to those reported for compost and its starting ingredients of manure, hay, and hardwood. For example, temporal dynamics of the composting process followed known patterns of ecological succession in herbivore manure [39]. The initial community was dominated by *Phycomycetes,* mostly *Mucorales,* such as *Mucor* and *Mortierella,* followed by ascomycota such as *Ascobolus* and *Chaetomium* spp., and finally basidiomycota such as *Coprinus* and *Stropharia* spp. [39]. Although manure- silage contained the greatest volume of manure, the other recipes also contained manure which explains why taxa commonly associated with animal feces (e.g., *Bacteroidetes, Firmicutes, γ-Proteobacteria, Chaetomium, Coprinis,* and *Ascobolus*) were found in all recipes [2], [19], [40]. *Zygomycota* in the *Harpelles, Mortierellales,* and *Mucorales* were associated more with manure-silage and hay than hardwood composts.

Fungi associated with tree bark were more commonly associated with hardwood compost, e.g., *Sordariomycetes* and *Agaricomycetes*. All recipes contained *Zopfiella* which typically arrives later in succession [41]. *Trichoderma, Alternaria,* and *Aspergillus* were not dominant on finished hardwood compost. To our knowledge, this is the first report of hardwood compost containing relatively abundant bacterial taxa within the *Acidobacteria* and *Verrucomicrobia* phyla, or of hay being a favorable habitat for *Actinobacteria* and *Gemmatimonadetes*. *Acidobacteria, Gemmatimonadetes,* and *Chloroflexi* have all been reported in waste water and sludge [42], [43], [44], but this is the first study to note their importance in the compost process. These taxa are notoriously difficult to culture and the ecological attributes of many members of these groups are not well-known.

6.4.2 COMPOST PROCESS

Different methods of composting after the thermophilic phase affect the dynamics and resulting composition of bacterial and fungal communities (Figure 2). We expected *Actinobacteria* and *Firmicutes* from windrow and ASP, and more *Chloroflexi, Acidobacteria, Bacteroidetes,* and *Gemmatimonadetes* in vermicompost [10]. The dominance of *Ascomycota* in both windrow and vermicompost processes has been documented in both culture-based [4] and 454-pyrosequencing [38] studies. However, in contrast to reports based on culturable fungi, we did not find *Fusarium* species to be a magnitude of order greater in windrow than vermicompost, or for *Trichoderma* species (anamorph: *Hypocrea, Hypocreaceae*) to be present exclusively in windrow [4].

Vermicompost had substantially different microbial communities when compared to those from ASP and windrow processes. These differences may be driven, in part, by differences in temperature regimes. Although vermicompost can be exclusively mesophilic, we inserted a thermophilic stage (ASP) prior to vermicomposting. We observed a greater diversity of bacteria in vermicompost than windrow and ASP (Figure 4). Our results support earlier reports that earthworms promote growth of bacteria [45] including *Bacteroidetes, Verrucomicrobia, Firmicutes*, and *Proteobacteria* [46]. There was also a trend for greater diversity of fungi in vermicompost and a relatively high abundance of fungi including *Mortierella* and *Arthrobotrys* supporting earlier reports [4], [47]. *Arthrobotrys* is a nematode-trapping fungus and been reported to be associated with earthworm inhabited soils [48]. In addition to the partially decomposed compost, regions of the digestive tract of earthworms are colonized by distinct communities of bacteria and fungi that may contribute to the overall microbial community of the compost product [49].

6.4.3 DYNAMICS OF COMPOSTING PROCESS

We know the thermophilic phase is important to reduce pathogen loads during composting (www.ams.usda.gov/nop). In contrast, the post-thermophilic phase is often ignored. Although recipe was held constant, microbial community composition diversified rapidly through time depending on whether the compost process was managed by windrow, ASP or vermicompost. To our knowledge, this is the first report of temporal dynamics in ASP or vermicompost.

Most of what we know about degradative succession in compost is from windrow studies. Relatively few taxa dominate during the thermophilic phase. For example, we observed that *Bacteroidetes* dominated at the end of thermophilic phase for windrow, supporting previous work [19]. In contrast, *Firmicutes* and γ-*Proteobacteria* were dominant after the thermophilic phase in ASP and vermicompost, respectively. All three of these bacterial phyla contain thermophiles [2]. Furthermore, relative abundance of γ-*Proteobacteria, Firmicutes* and *Actinobactera* are reported as indicators of disease suppression [50].

The fungi we observed at the end of thermophilic phase were similar to those reported previously from compost. For example, *Dothideomycetes* and *Eurotiales* were both abundant after the thermophilic phase [2], [15]. In contrast to Anastasi et al. [4], we found *Taloromyces* in compost made by all three processes rather than just vermicompost. Other fungi that were relatively abundant after the thermophilic phase have been reported for composts containing ingredients similar to our study. For example, *Cladosporium* has been reported previously on compost based on cattle manure [2]. *Thermomyces* and *Penicillium* have both been isolated from composts containing hardwood bark and manure [2].

Near the end of the composting process, a different and more complex community develops that includes chytrids, protists, *Ascomycota*, and *Stamenopiles* [15]. In windrow, *Chloroflexi* and γ-*Proteobacteria* decreased in relative abundance during the cooling and curing phases and were surpassed in abundance by *Bacteroidetes* in finished composts. This supports earlier reports that *Bacteroidetes* are relatively abundant, and more abundant than γ-*Proteobacteria*, in finished compost [10], [16]. In contrast, *Chloroflexi* decreased and abundances of α-*Proteobacteria* were half those of γ-*Proteobacteria* in the final product [23]. *Actinobacteria* become more abundant during the curing phase [51], [52], [53], [54]. *Actinobacteria* are less likely to be found in vermicompost with the more active microbial communities promoted by earthworms [47].

Numerous mesophilic fungi proliferate during the cooling and curing phases [50]. Many of the fungi we found are known to be widespread saprophytes on soil and dead plant tissue [55]. Similar to Anastasi et al. [4], *Acremonium* and *Cladosporium* occurred in both windrow and vermicompost, and *Scytalidium* was more abundant in windrow than vermicompost. It is no surprise that *Arthrographis* and *Galactomyces* occur in compost given their ability to produce cellulolytic enzymes [56], [57], however, this is the first report for composts made by ASP and vermicompost processes, specifically.

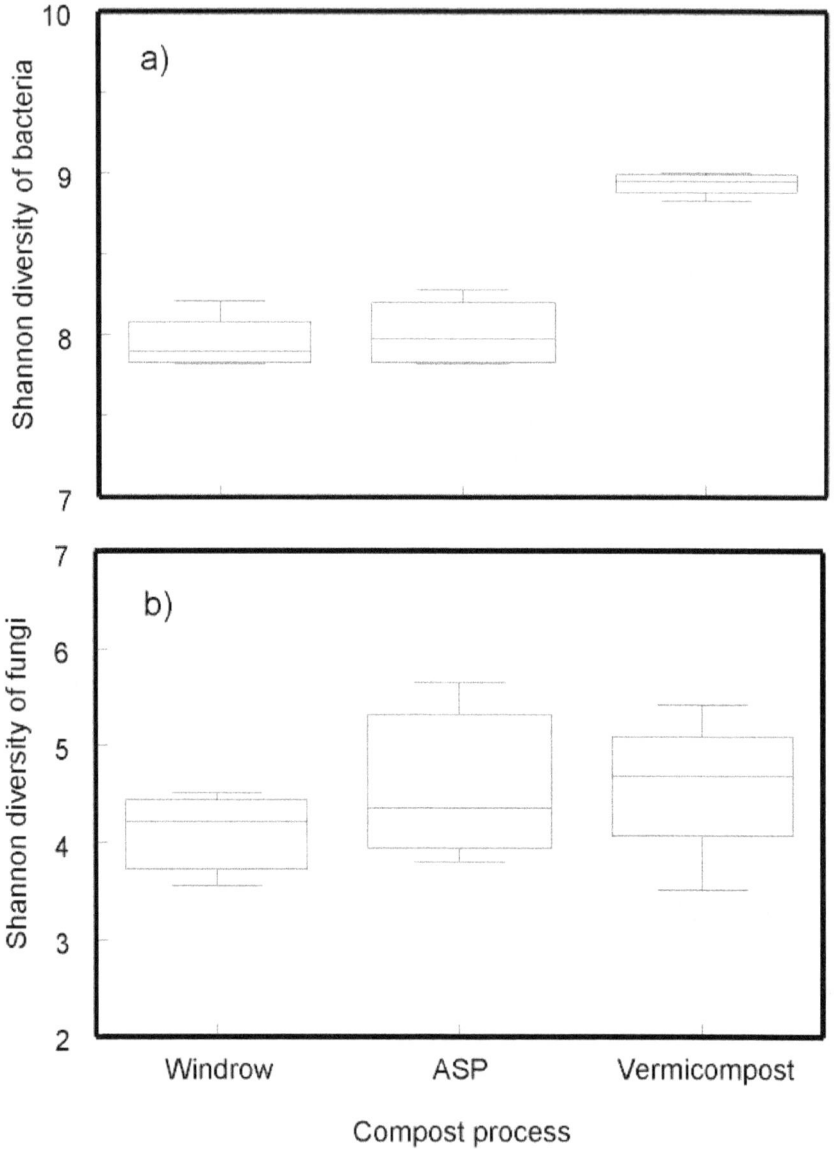

FIGURE 4: Shannon diversity of a) bacteria and b) fungal communities within a standardized recipe finished by windrow, aerated static pile (ASP) or vermicompost. Shannon diversity is computed as $H' = -\Sigma(pi \ln pi)$ where p represents the proportion of taxon i in the community. Box-whisker plots are illustrated.

6.5 CONCLUSION

Microbial communities are abundant and diverse in compost. Communities are organized and influenced by recipe, and post-thermophilic treatment. Composition starts similarly after thermophilic phase and shifts dynamically through time. Economic considerations have driven commercial composters to expedite the composting process. As a result, the focus has been on the effectiveness of the thermophilic phase. The curing phase offers a substrate and climate conducive for microbial recolonization which can be accomplished either by inoculating post-thermophilic compost or preparing a palatable substrate that provides a competitive advantage for colonization by bacteria and fungi that offer biological control, slow-release fertility, and/or promote plant growth. Future research can build on the microbial results presented here to determine which recipe and post-thermophilic phase are best to achieve desired agricultural goals of weed management, disease suppression, and plant growth promotion.

REFERENCES

1. US-EPA (2011) Municipal solid waste generation, recycling, and disposal in the United States Tables and Figures for 2010. Available: http://www.epa.gov/osw/non-haz/municipal/msw99.htm. Accessed May 10, 2013.
2. Ryckeboer J, Mergaert J, Vaes K, Klammer S, de Clercq D, et al. (2003) A survey of bacteria and fungi occurring during composting and self-heating processes. Ann Microbiol 53: 349–410.
3. Ahmad R, Jilani G, Arshad M, Zahir ZA, Khalid A (2007) Bio-conversion of organic wastes for their recycling in agriculture: an overview of perspectives and prospects. Ann Microbiol 57: 471–479. doi: 10.1007/bf03175343
4. Anastasi A, Varese GC, Marchisio VF (2005) Isolation and identification of fungal communities in compost and vermicompost. Mycologia 97: 33–44. doi: 10.3852/mycologia.97.1.33
5. Ashraf R, Shahid F, Ali TA (2007) Association of fungi, bacteria and actinomycetes with different composts. Pak J Bot 39: 2141–2141.
6. Franke-Whittle IH, Knapp BA, Fuchs J, Kaufmann R, Insam H (2009) Application of COMPOCHIP microarray to investigate the bacterial communities of different composts. Microbial Ecol 57: 510–521. doi: 10.1007/s00248-008-9435-2
7. Green SJ, Michel FC, Hadar Y, Minz D (2004) Similarity of bacterial communities in sawdust- and straw-amended cow manure composts. FEMS Microbiol Lett 233: 115–123. doi: 10.1016/j.femsle.2004.01.049

8. Hill GT, Mitkowski NA, Aldrich-Wolfe L, Emele LR, Jurkonie DD, et al. (2000) Methods for assessing the composition and diversity of soil microbial communities. Appl Soil Ecol 15: 25–36. doi: 10.1016/s0929-1393(00)00069-x

9. Dees PM, Ghiorse WC (2001) Microbial diversity in hot synthetic compost as revealed by PCR-amplified rRNA sequences from cultivated isolates and extracted DNA. FEMS Microbiol Ecol 35: 207–216. doi: 10.1111/j.1574-6941.2001.tb00805.x

10. Fracchia L, Dohrmann A, Martinotti M, Tebbe C (2006) Bacterial diversity in a finished compost and vermicompost: differences revealed by cultivation-independent analyses of PCR-amplified 16S rRNA genes. Appl Microbiol Biot 71: 942–952. doi: 10.1007/s00253-005-0228-y

11. Hansgate AM, Schloss PD, Hay AG, Walker LP (2005) Molecular characterization of fungal, community dynamics in the initial stages of composting. FEMS Microbiol Ecol 51: 209–214. doi: 10.1016/j.femsec.2004.08.009

12. Guo Y, Zhang J, Deng C, Zhu N (2012) Spatial heterogeneity of bacteria: Evidence from hot composts by culture-independent analysis. Asian Austral J Anim 25: 1045–1054. doi: 10.5713/ajas.2011.11341

13. Nakasaki K, Nag K, Karita S (2005) Microbial succession associated with organic matter decomposition during thermophilic composting of organic waste. Waste Manage Res 23: 48–56. doi: 10.1177/0734242x05049771

14. Peters S, Koschinsky S, Schwieger F, Tebbe CC (2000) Succession of microbial communities during hot composting as detected by PCR–single-strand-conformation polymorphism-based genetic profiles of small-subunit rRNA genes. Appl Environ Microbiol 66: 930–936. doi: 10.1128/aem.66.3.930-936.2000

15. Bonito G, Isikhuemhen OS, Vilgalys R (2010) Identification of fungi associated with municipal compost using DNA-based techniques. Biorsource Technol 101: 1021–1027. doi: 10.1016/j.biortech.2009.08.109

16. Danon M, Franke-Whittle IH, Insam H, Chen Y, Hadar Y (2008) Molecular analysis of bacterial community succession during prolonged compost curing. FEMS Microbiol Ecol 65: 133–144. doi: 10.1111/j.1574-6941.2008.00506.x

17. Hultman J, Vasara T, Partanen P, Kurola J, Kontro MH, et al. (2010) Determination of fungal succession during municipal solid waste composting using a cloning-based analysis. J Appl Microbiol 108: 472–487. doi: 10.1111/j.1365-2672.2009.04439.x

18. Klamer M, Baath E (1998) Microbial community dynamics during composting of straw material studied using phospholipid fatty acid analysis. FEMS Microbiol Ecol 27: 9–20. doi: 10.1111/j.1574-6941.1998.tb00521.x

19. Sasaki H, Nonaka J, Otawa K, Kitazume O, Asano R, et al. (2009) Analysis of the structure of the bacterial community in the livestock manure-based composting process. Asian Austral J Anim 22: 113–118.

20. Tiquia SM (2005) Microbial community dynamics in manure composts based on 16S and 18S rDNA T-RFLP profiles. Environ Technol 26: 1101–1114. doi: 10.1080/09593332608618482

21. Viera B, Madayanti VE, Aryantha F, Akhmaloka INP (2012) Succession of eukaryotic communities during traditional composting of domestic waste based on PCR-DGGE analysis. J Pure Appl Microbiol 6: 525–536.

22. Alfreider A, Peters S, Tebbe CC, Rangger A, Insam H (2002) Microbial community dynamics during composting of organic matter as determined by 16S ribosomal DNA analysis. Compost Sci Util 10: 303–312. doi: 10.1080/1065657x.2002.10702094

23. de Gannes VD, Eudoxie G, Hickey WJ (2013a) Prokaryotic successions and diversity in composts as revealed by 454-pyrosequencing. Bioresource Technol 133: 573–580. doi: 10.1016/j.biortech.2013.01.138

24. Fernandez-Gomez MJ, Nogales R, Insam H, Robero E, Goberna M (2012) Use of DGGE and COMPOCHIP for investigating bacterial communities of various vermicomposts produced from different wastes under dissimilar conditions. Sci Total Environ 414: 664–671. doi: 10.1016/j.scitotenv.2011.11.045

25. Partanen P, Hultman J, Paulin L, Auvinen P, Romantschuk M (2010) Bacterial diversity at different stages of the composting process. BMC Microbiol 10: 94 Available: http://www.biomedcentral.com/1471-2180/10/94. Accessed 25 January 2013.

26. Lauber CL, Hamady M, Knight R, Fierer N (2009) Pyrosequencing-based assessment of soil pH as a predictor of soil bacterial community structure at the continental scale. Appl Environ Microbiol 75: 5111–5120. doi: 10.1128/aem.00335-09

27. Fierer N, Leff JW, Adams BJ, Nielsen UN, Bates ST, et al. (2012) Cross-biome metagenomic analysis of soil microbial communities and their functional attributes. PNAS doi 10.1073pnas.1215210110/pnas.1215210110.

28. Caporaso JG, Lauber CL, Walters WA, Berg-Lyons D, Lozupone CA, et al. (2012) Ultra-high-throughput microbial community analysis on the Illumina HiSeq and MiSeq platforms. The ISME J 6: 1621–1624. doi: 10.1038/ismej.2012.8

29. Gardes M, Bruns TD (1993) ITS primers with enhanced specificity for basidiomycetes – application to the identification of mcorrhizae and rusts. Mol Ecol 2: 113–118. doi: 10.1111/j.1365-294x.1993.tb00005.x

30. McGuire KL, Payne SG, Palmer MI, Gillikin CM, Keefe D, et al. (2013) Digging the New York City skyline: Soil fungal communities in green roofs and city parks. PLoS One 8: e58020. doi: 10.1371/journal.pone.0058020

31. Caporaso JG, Kuczynski J, Stombaugh J, Bittinger K, Bushman FD, et al. (2010) QIIME allows integration and analysis of high-throughput community sequencing data. Nat Methods 7: 335–336. doi: 10.1038/nmeth.f.303

32. Edgar RC (2010) Search and clustering orders of magnitude faster than BLAST. Bioinformatics 26: 2460–2461. doi: 10.1093/bioinformatics/btq461

33. McDonald D, Price MN, Goodrich J, Nawrocki EP, DeSantis TZ, et al. (2011) An improved Greengenes taxonomy with explicit ranks for ecological and evolutionary analyses of bacteria and archaea. ISME J 6: 610–618. doi: 10.1038/ismej.2011.139

34. Bates S, Ahrendt S, Bik H, Bruns T, Caporaso J, Cole J, et al. (2013) Meeting Report: Fungal ITS Workshop (October 2012). Standards In Genomic Sciences 8 doi:10.4056/sigs.3737409.

35. Wang Q, Garrity GM, Tiedje JM, Cole JR (2007) Naive Bayesian classifier for rapid assignment of rRNA sequences into the new bacterial taxonomy. Appl Environ Microbiol 73: 5261–5267. doi: 10.1128/aem.00062-07

36. Clarke KR, Gorley RN (2001) Plymouth Routines in Multivariate Ecological Research (Primer) Version 5: User Manual/Tutorial. Primer-E Ltd, Plymouth, UK.

37. R Core Team (2012) R: A language and environment for statistical computing. R Foundation for Statistical Computing, Vienna, Austria. Available: http://www.R-project.org/. Accessed 7 January 2013.

38. de Gannes VD, Eudoxie G, Hickey WJ (2013b) Insights into fungal communities in composts revealed by 454-pyrosequencing: implications for human health and safety. Front Microbiol doi:10.3389/fmicb.2013.00164.

39. Hudson HJ (1968) The ecology of fungi on plant remains above the soil. New Phytol 67: 837–874. doi: 10.1111/j.1469-8137.1968.tb06399.x

40. Richardson MJ (2001) Diversity and occurrence of coprophilous fungi. Mycol Res 105: 387–402. doi: 10.1017/s0953756201003884

41. Pillinger JM, Cooper JA, Ridge I, Barrett PRF (1992) Barley straw as an inhibitor of algal growth III: the role of fungal decomposition. J Appl Phycol 4: 353–355. doi: 10.1007/bf02185793

42. Bjornsson L, Hugenholtz P, Tyson GW, Blackall LL (2002) Filamentous *Chloroflexi* (green non-sulfur bacteria) are abundant in wastewater treatment processes with biological nutrient removal. Microbiol-Sgm 148: 2309–2318.

43. Quaiser A, Ochsenreiter T, Lanz C, Schuster SC, Treusch AH, et al. (2003) *Acidobacteria* form a coherent but highly diverse group within the bacterial domain: evidence from environmental genomics. Mol Microbiol 50: 563–575. doi: 10.1046/j.1365-2958.2003.03707.x

44. Zhang HJ, Sekiguchi Y, Hanada S, Hugenholtz P, Kim H, et al. (2003) Gemmatimonas aurantiaca gen. nov., sp. nov., a gram-negative, aerobic, polyphosphate-accumulating micro-organism, the first cultured representative of the new bacterial phylum Gemmatimonadetes phyl. nov. Int J Syst Evol Microbiol 53: 1155–1163. doi: 10.1099/ijs.0.02520-0

45. Yakushev AV, Bubnov IA, Semenov AM (2011) Estimation of the effects of earthworms and initial substrates on the bacterial community in vermicomposts. Eurasian Soil Sci 44: 1117–1124. doi: 10.1134/s1064229311100164

46. Bernard L, Chapuis-Lardy L, Razafimbelo T, Razafindrakoto M, Pablo AL, et al. (2012) Endogeic earthworms shape bacterial functional communities and affect organic matter mineralization in a tropical soil. ISME J 6: 213–222. doi: 10.1038/ismej.2011.87

47. Lazcano C, Gomez-Brandon M, Dominguez J (2008) Comparison of the effectiveness of composting and vermicomposting for the biological stabilization of cattle manure. Chemosphere 72: 1013–1019. doi: 10.1016/j.chemosphere.2008.04.016

48. Sukhjeet K, Kaul VK (2006) Evaluation of substrates for growth of nematophagous fungus, Arthrobotrys oligospora. Ann Plant Prot Sci 14: 456–458.

49. Byzov BA, Nechitaylo TY, Burmazhkin BK, Kurakov AV, Golyshin PN, et al. (2009) Culturable microorganisms from the earthworm digestive tract. Microbiology 78: 360–368. doi: 10.1134/s0026261709030151

50. Hadar Y, Papadopoulou KK (2012) Suppressive composts: microbial ecology links between abiotic environments and healthy plants. Annu Rev Phytopathol 50: 133–153. doi: 10.1146/annurev-phyto-081211-172914

51. Cahyani VR, Matsuya K, Asakawa S, Kimura M (2003) Succession and phylogenetic composition of bacterial communities responsible for the composting process

of rice straw estimated by PCR-DGGE analysis. Soil Sci Plant Nutr 49: 619–630. doi: 10.1080/00380768.2003.10410052

52. Steger K, Jarvis Å, Vasara T, Romantschuk M, Sundh I (2007) Effects of differing temperature management on development of Actinobacteria populations during composting. Res Microbiol 158: 617–624. doi: 10.1016/j.resmic.2007.05.006

53. Tang JC, Kanamori T, Inoue Y, Yasuta T, Yoshida S, et al. (2004) Changes in the microbial community structure during thermophilic composting of manure as detected by the quinone profile method. Process Biochem 39: 1999–2006. doi: 10.1016/j.procbio.2003.09.029

54. Xiao Y, Zeng GM, Yang ZH, Ma YH, Huang C, et al. (2011) Effects of continuous thermophilic composting (CTC) on bacterial community in the active composting process. Microb Ecol 62: 599–608. doi: 10.1007/s00248-011-9882-z

55. Kirk PM, Cannon PF, Minter DW, Stalpers JA (2008) Ainsworth & Bisby's Dictionary of the Fungi, Tenth Edition. Wallingsford: CABI. 784 p.

56. Eida MF, Nagaoka T, Wasaki J, Kouno K (2011) Evaluation of cellulolytic and hemicellulolytic abilities of fungi isolated from coffee residue and sawdust composts. Microbes Environ 28: 220–227. doi: 10.1264/jsme2.me10210

57. Okeke BC, Obi SKC (1993) Production of cellulolytic and xylanolytic enzymes by an *Arthrographis* species. World J Microb Biot 9: 345–349. doi: 10.1007/bf00383077

There is one table and several supplemental files that are not available in this version of the article. To view this additional information, please use the citation on the first page of this chapter.

CHAPTER 7

Effects of Bulking Agents, Load Size or Starter Cultures in Kitchen-Waste Composting

NORAZLIN ABDULLAH, NYUK LING CHIN, MOHD NORIZNAN MOKHTAR, and FARAH SALEENA TAIP

7.1 INTRODUCTION

Disposal of kitchen waste, which contains about 80% of moisture to the landfills, causes various problems like easy putrefaction, offensive odour and pollution of ground and surface water by leachate (Rogoshewski et al. 1983; Wang et al. 2001). Due to interruption of the carbon cycle by disposal of waste to landfills, organic kitchen waste requires proper composting system to reduce its uncontrolled degradation on disposal sites and subsequent greenhouse gases, odour and nutrient emissions (Luostarinen and Rintala 2007). In addition, kitchen waste may be wasted if it is just dumped to landfills as it will break up naturally and never be used directly again, having its nutritious matter lost within the waste. Therefore, a

Effects of Bulking Agents, Load Size or Starter Cultures in Kitchen-Waste Compostin © Abdullah N, Chin NL, Mokhtar MN, and Taip FS. International Journal of Recycling of Organic Waste in Agriculture *2,3 (2013), doi:10.1186/2251-7715-2-3. Licensed under Creative Commons Attribution 2.0 Generic License, http://creativecommons.org/licenses/by/2.0/.*

kitchen waste composter was designed to allow the composting process to be conducted in a consistent manner.

While people give attention to recycled inorganic wastes such as plastics, glass and metals, kitchen waste which is rich in organic material and possesses more than 90% of biodegradability can be easily recycled into compost (Veeken and Hamelers 1999). Composting of kitchen waste can be an effective method to reduce waste in landfills while helping conserve the environment. As kitchen waste is produced everyday and everywhere from processed and unprocessed food for human consumption, its composition is quite variable. An optimised kitchen waste formulation and composition involving the use of bulking materials, waste load size and presence of microbes are important in ensuring the commencement of an effective composting process (Fang et al. 2001; Ishii and Takii 2003; Cekmecelioglu et al. 2005; Stabnikova et al. 2005; Cayuela et al. 2006; Chang and Hsu 2008).

Recognised bulking agents used for composting include sawdust (Sundberg and Jönsson 2005; Kalamdhad et al. 2008), rice hulls and chips of tree cuttings (Chikae et al. 2006), horticultural waste compost (Stabnikova et al. 2005) and mulch hay and wood shavings (Cekmecelioglu et al. 2005). Newspaper contains 9% of moisture content (MC) (Wayman et al. 1992) and 94% of volatile solids with lignin content of the volatile solids ranging from 16% to 22% (Sun and Cheng 2002), while the onion peels contain about 54% of volatile solids (Lubberding et al. 1988) with 1.5% dry weight of lignin (Suutarinen et al. 2003). Both can be suitable bulking materials in the composting of kitchen waste as they contain high carbon contents (Abdullah and Chin 2010). Newspaper is distributed throughout the land, in practically every house and building. The Malaysian Audit Bureau of Circulations reports that newspaper circulations have increased from 2005 to 2009, where the total of publications for the last 6 months of 2009 was 4,100,486 and 576,663 for West and East Malaysia, respectively. Onions are one of the major vegetables consumed in Malaysia (Kamil et al. 2010) and onion peels are the most common waste disposed in almost every kitchen. Among the three important constituents of plant cell wall material, the cellulose, lignin and hemicellulose, lignin is particularly difficult to biodegrade and reduces the bioavailability of the other cell wall constituents (Naik et al. 2012). The domination of different indigenous microorganism population at various stages of composting plays a distinct role in degrading lignin (Raut et al. 2008; Belyaeva and Haynes 2009; Huang et al. 2010).

Waste load size influences the temperature development during a composting process (Cayuela et al. 2006). It affects the achievement of the thermophilic phase where it is an essential stage for sanitation and killing all pathogens. Small heap sizes may not produce enough heat to reach the thermophilic phase, while too large heap sizes may prevent air passage into the centre which will paralyse the composting process (Cayuela et al. 2006). The windrow system requires a large heap to achieve the thermophilic stage. This fact is supported by Cekmecelioglu et al. (2005) who performed a composting process using about 11-m long, 2.5-m wide and 1.2-m high heap using conventional layering and mixing methods. Elsewhere, Sellami et al. (2008) have developed cone-shaped heap of 2-m high with a 3-m wide base for co-composting of exhausted olive cake, poultry manure and sesame bark. The other composting method using the in-vessel system requires a closed container, which is also known as a reactor to perform the composting process in small volume. This system can be placed inside or outside the building, as long as the substrates are protected from environmental effects and the process is under controlled conditions. Ishii and Takii (2003) found that smaller stacks are better for food-waste composting. A smaller scale of food waste progresses faster as food waste contains much more easily degradable compounds, and it takes a shorter time to turn into compost.

The application of starter culture has been found advantageous in composting especially for cellulosic waste, which is difficult to be broken down into smaller pieces (Volchatova et al. 2002). Its use is not a new practice for composting, where Wang et al. (2003) added a starter bacterial culture of *Bacillus thermoamylovorans* SW25 in composting mixtures of anaerobic dewatered sewage sludge and vegetable food waste. Although Fang et al. (2001) suggested that starter culture might assist in the composting process, Stabnikova et al. (2005) found that starter cultures are useless in food waste aerobic thermophilic bioconversion as sufficient air and thermophilic temperature ensure a quick composting process. The starter culture governs the process by reducing the accumulation of actinomycetes and fungi, which discharge allergenic spores into the air, and lowering the risk of harmful microorganism growth in the end product (Wang et al. 2003). Previous researchers have used *B. thermoamylovorans*

SW25 (Wang et al. 2003; Stabnikova et al. 2005), cultures of three *Bacillus* species (such as *B. brevis, B. coagulans* and *B. licheniformis* (Fang et al. 2001)), and a specially developed association of microorganism (Volchatova et al. 2002) in their composting studies.

The objectives of this work are to study the effects of two common bulking media, newspaper or onion peels with compost load sizes of 2 and 6 kg and the use of starter culture in kitchen-waste composting consisting of vegetable scraps and fish processing waste using an optimised formulation using the in-vessel system.

TABLE 1: Percentage (%) of moisture, carbon and nitrogen content, and CN ratio (Abdullah and Chin 2010)

Substrates	Moisture content (%)	Carbon content (%)	Nitrogen content (%)	CN ratio
Vegetable scraps	93.32±0.33	37.87±1.14	2.94±0.14	12.87±0.24
Fish processing waste	75.83±0.43	42.60±1.73	8.14±1.24	5.30±0.61
Newspaper	6.50±0.32	50.56±0.02	0.18±0.01	288.15±13.36
Onion peels	11.69±0.22	50.27±0.63	0.83±0.01	60.91±1.82

Percentage and CN ratio are in mean and standard deviation.

7.2 METHODS

7.2.1 COMPOSTING MATERIAL PREPARATION

The main composting substrates were vegetable scraps from spinach (*Spinacia oleracea* L.) and fish processing waste of Indian mackerels (*Rastrelliger kanagurta* C.). Newspaper and onion peels were chosen as they are easily available bulking agents, can act as moisture adjuster due to their low moisture content and have high cellulose content, which can be a good source of carbon. The characteristics of both waste and bulking agents in terms of its percentage of moisture, carbon and nitrogen contents, and the values of the carbon-to-nitrogen (CN) ratio have been measured in an earlier study (Abdullah and Chin 2010) and are given in Table 1.

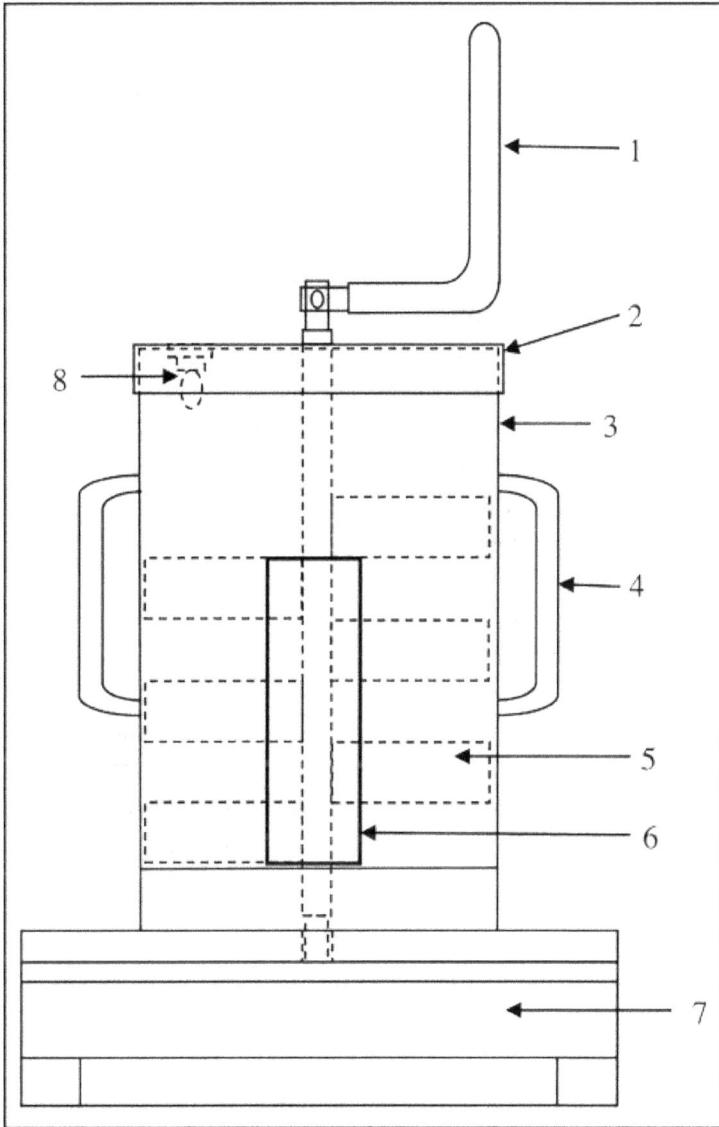

FIGURE 1: Schematic diagram of the designed kitchen waste composter. 1 = Turner's handle, 2 = perforated cover, 3 = body, 4 = handle, 5 = blade, 6 = window, 7 = tray, 8 = light.

7.2.2 KITCHEN WASTE COMPOSTER DESIGN

As kitchen waste which contains very high moisture is difficult to be self-heated up to 45°C, two self-designed kitchen waste composters which included features for temperature and moisture content control were fabricated to allow parallel experiments to be conducted consistently (Figure 1). The kitchen composter consists of three parts, which are the perforated cover, the body and a collector of water and end product. All parts were made from stainless steel 304, except for its transparent window made of polycarbonate that is a heat-resistant plastic for convenient viewing of the waste inside the composter. The external part of body and the perforated cover were insulated using cloth to minimise heat loss. A 25-W bulb was attached to the perforated cover to provide heat internally. The opening at the top allows feeding of fresh waste, and the small holes in the perforated cover allow air movement to maintain an aerobic process. The long L-shaped handle on the top is for manual turning of the mixer blade to aerate and homogenise the sample. The blades were slanted 5° in opposite directions to keep a good blending by moving the mixture up and down. The two side handles at the body of the composter aid the rotation of the body up to 360° when collecting end products. A collector tray was inserted below the body to collect the leachate water or end product. Between the body and the collector, there was a medium hole to collect the end product and many tiny holes to allow excessive water draining.

7.2.3 COMPOSTING AND SAMPLING

The optimised mixture formulation for kitchen-waste composting was referred from Abdullah and Chin (2010) following the fixed CN ratio of 30 and MC of 60%. Table 2 shows the amounts of substrate compositions for investigations on bulking agents, load size and use of starter culture. All substrates were weighed using a scale (SSB12001, Mettler Toledo, Switzerland), then mixed manually in a basin before loading into the composter. The composting was conducted simultaneously using two composters for each experiment. The first experiment was the comparison of the effectiveness of the bulking media, i.e. between the newspaper and onion peels at 2-kg load size. After finding that the onion peels were more effective in composting, they were used at two load sizes, 2 and 6 kg, before further investigation on the effect of presence of starter cultures using 6 kg of kitchen waste. Each compost cycle was for 30-days, and the composters were placed in a laboratory at $27 \pm 5°C$. The composts were also turned and mixed in the composters each time after sampling to prevent shortage of oxygen pore required for aerobic composting. The pipe water sprayed onto the composting substrates inside the composters to maintain an MC of 60% was left overnight before use to reduce its chlorine content. Samples were taken on days 0, 1, 4, 8, 10, 16, 23 and 30 of the composting period. The bulb in the composter was switched on to help the composting material to achieve the thermophilic temperature of 45°C and until the compost temperature stabilised towards the end of the composting process where the CN ratio reduced to less than 15 or the compost reached the matured stage.

TABLE 2: Amount of substrate compositions used in each experiment

Substrates	Amount (g)	
	Composter 1	Composter 2
Bulking agents	Newspaper	Onion peels
Vegetable scraps	971	880
Fish processing waste	355	395
Newspaper	675	0
Onion peels	0	725
Load size	2 kg	6 kg
Vegetable scraps	880	2,641
Fish processing waste	395	1,185
Onion peels	725	2,174
Starter culture	With	Without (control)
Vegetable scraps	2,641	2,641
Fish processing waste	1,185	1,185
Onion peels	2,174	2,174

For composting with added starter culture, a solution of bacteria from *Aeromonas* sp., *Azotobacter* sp., *Bacillus* sp., *Clostridium* sp., *Pseudomonas* sp., *Thermomonaspora* sp. and *Trichurus* sp., and eight plates of fungi from *Aspergillus* sp., *Cellulomonas* sp., *Chaetomium* sp., *Coprinus* sp., *Microbispora* sp., *Penicillium* sp., *Thermoactinomyces* sp. and *Trichoderma* sp. were provided by the Biotechnology Research Centre, Malaysian Agricultural Research and Development Institute. This microbe cocktail was formulated after considering the functions of each microorganism isolated from soil. The *Bacillus* sp., *Clostridium* sp., *Pseudomonas* sp., *Aspergillus* sp., *Cellulomonas* sp., *Chaetomium* sp., *Penicillium* sp. and *Trichoderma* sp. play important roles in degrading cellulose and other carbohydrates (Bhatt and Kausadikar 2010). Proteins are degraded to individual amino acids mainly by fungi, *Actinomycetes, Bacillus* sp., *Clostridium* sp. and *Pseudomonas* sp. (Bhatt and Kausadikar 2010). Bacteria like the *Pseudomonas* sp., *Clostridium* sp. and *Bacillus* sp. and fungi like *Trichoderma* sp., *Penicillium* sp. and *Aspergillus* sp. help degrade toxic chemicals and pesticides into non-toxic substances, thus minimising any damage caused by harmful chemicals to the ecosystem (Bhatt and Kausadikar 2010). Both bacteria and fungi were mixed using a kitchen blender (PB-323 T, Pensonic, Malaysia) before being sprayed onto the waste in composter 1 on day 1 of composting. This is to allow the waste inside the kitchen waste composter to stabilise and so that the starter culture could easily adapt to the surroundings inside the kitchen waste composter and perform good composting activities.

7.2.4 ANALYTICAL METHODS

During the 30-days of kitchen-waste composting, the progression of temperature and MC (Abdullah and Chin 2010) were determined. The temperature was measured at 50% depth from the surface of compost materials at four different positions. In assessing the maturity level of the compost, the CN ratio was determined. The content of the volatile solids was calculated after measuring the ash content using the dry ashing method to obtain the total organic carbon (TOC) content (Abdullah and Chin 2010). The total nitrogen content was measured using the micro-Kjeldahl method

(Mohee et al. 2008; Unmar and Mohee 2008), where the digested sample was distilled with 45% sodium hydroxide and 2% boric acid, and titrated with 0.05 N sulphuric acid until neutral; the same procedure was repeated for the blank sample. The CN ratio was calculated by dividing the TOC content with the total nitrogen content. All the analyses were conducted on samples collected using a standard sampling method, except for temperature measurements which were performed in situ.

7.2.5 STATISTICAL ANALYSES

All samples were analysed in triplicates, except for the temperature measurements, which were repeated four times. The averages and standard deviation of the means as the error bars were calculated using Microsoft Excel 2007 (Vista Edition, Microsoft Corporation, USA). Statistical analyses were made using statistical analysis software (SAS 9.2, SAS Institute, Inc., USA). The data were subjected to one-way analysis of variance (ANOVA), generalised linear model (GLM) and Duncan's multiple range tests (DMRT) at alpha level, $\alpha = 0.05$. ANOVA was implemented to all parameters, except for those data with unbalanced sample size and missing data using the GLM statistical analyses. Where significant differences were obtained giving $p < 0.05$, individual means were tested using the DMRT to compare the significant difference between the two treatments means, i.e. newspaper blend with the onion peels blend, 2 kg with 6 kg of load size and starter culture with control.

7.3 RESULTS AND DISCUSSION

In general, the MC of the compost increased with time due to vaporisation and the trapped moisture loss inside the composter. The values were still acceptable as long as the air supply in the compost mass is sufficient in keeping the microbes alive (Unmar and Mohee 2008). The pH of the waste mixture for all treatments increased up to 9 from acidic conditions through the 30-days of composting, while the electrical conductivity (EC) increased and was in the range of 3 to 6 dS/m. The EC was low at the thermophilic temperature because a high amount of nutrients that were ingested by the microorganisms made the nutrients insoluble in water and consequently produced a low salinity. With colour measurements, the blend with onion peels became black more rapidly than the blend with newspaper ($p < 0.0001$), showing the colour of matured and stable compost. The waste compost colour changes to black in the 2- and 6-kg load sizes, and both mixtures with and without starter culture were not significantly different with $p = 0.2031$ and $p = 0.8383$, respectively. There were no clear trends in the all microbial numbers due to a diverse microbial community in the organic waste, which comply with the reports by Sundberg et al. (2004) who believed that the temperatures affected the active microbial numbers.

7.3.1 TEMPERATURE PROFILES

Figure 2 illustrates the temperature profiles showing typical phases of a composting process for investigations using bulking agents of newspaper and onion peels, investigations of load sizes at 2 and 6 kg, and use of starter culture during a 30-day kitchen-waste composting process. The temperature profiles displayed quite similar patterns for all composting batches where it increased at the early stage before decreasing gradually. For the first 24 h in Figure 2a, the compost with newspaper was self-heated from 26.7°C to about 39.2°C, while the blend with the onion peels was self-heated from 30.3°C to about 44.9°C. The temperature change in onion peels blend during composting followed a typical pattern displayed by many other composting system, i.e. organic fraction of municipal solid wastes, raw sludge, anaerobically digested sludge, animal by-products from slaughterhouses and partially hydrolysed hair from the leather industry (Pagans et al. 2006), co-composting of exhausted olive-cake with poultry manure and sesame shells (Sellami et al. 2008), green leaves, green branches, grass, brown branches and brown leaves Unmar and Mohee (2008), and co-composting of green tea waste and rice bran (Khan et al. 2009). The newspapers, however, did not assist the compost to achieve thermophilic temperature of 45°C. The temperature of the compost with the onion peels rose to 51.3°C on day 4 and dropped sharply to 44.1°C on day 7. It stayed in the thermophilic phase for about 3 days. The temperature decrease is probably due to the excessive loss of compost volume, and when the compost moves into the cool phase, it produces stable compost. The average temperature of the two composts with different bulking agents, newspaper and onion peels was significantly different at $p < 0.0001$ when compared with ambient temperatures. The DMRT resulted greater mean of temperature in compost with onion peels, followed by compost with newspaper and ambient temperature.

Both compost load sizes at 2 and 6 kg using onion peels as shown in Figure 2b present a quite similar temperature trend with compost using food waste, manure and bulking agent mixture by Cekmecelioglu et al. (2005). The composting started with mesophilic temperature, continued to thermophilic temperature and then drop to ambient temperature. During the first 24 h of composting, the temperature of the waste rose 17.5°C for the 2-kg load and 8°C for the 6 kg. The 6-kg load contained more carbon to be degraded compared to the 2 kg; thus, it was observed that the 2-kg load has entered the mesophilic stage 10 days earlier than the 6 kg, on day 6 and 16, respectively. The temperature fluctuations observed during the composting process were because of the turning and watering activities. The 6-kg load size has a higher mean of temperature followed by 2-kg load size and ambient temperature.

Figure 2c shows that the temperatures of the composts with added starter culture and control increased quickly to thermophilic temperatures of 53°C and 51.1°C, respectively, on day 3 of composting due to the compost load size. The presence of an active microbial community also helps start the degradation process immediately and reduces the particle size (Sundberg and Jönsson 2005). The thermophilic phase of both composts lasted for a long period of 17 days, probably due to the availability of easily degradable organic matter, energetic compounds (like protein) and large amounts of organic nitrogen in the waste (Pagans et al. 2006). With a large amount of waste, the high thermophilic temperatures worked on breaking down the proteins, fats and complex carbohydrates like cellulose and hemicelluloses. The compost with added starter culture achieved the highest temperature of 54.3°C on day 8, which is slightly higher and one day earlier than the control (53.8°C on day 9). The temperature of the mixtures began to decline on day 8 for mixture with starter culture and day 9 for control, probably due to convective loss (Palmisano et al. 1993) and a higher amount of readily degradable carbon. At the end of the process, heat was released progressively, causing the temperature for both mixtures to decrease and tended to meet the ambient temperature. This tendency was found with composting fish offal in reactors (Laos et al. 2002) and composting of green tea waste and rice bran (Khan et al. 2009), which implied that the rapidly degradable organic matter had been reduced (Sundberg and Jönsson 2005). The average temperatures of the compost with added starter culture and the control were not significantly different from the DMRT results.

7.3.2 DEGRADATION RATE PROGRESSION

Figure 3 illustrates the decreasing trends of volatile solids content during the 30-day composting period. All three figures were comparable with the findings of Cekmecelioglu et al. (2005) in terms of volatile solids content. Unmar and Mohee (2008) considered that the volatile solids are a good indicator of how biological degradation occurred over time. Figure 3a shows that the initial volatile solids contents for both blends (onion peels and newspaper) were the same at 85%. The final volatile solids in compost with onion peels then reduced to 60.9% compared to the newspaper at 76.1%, indicating that onion peel compost had a higher degradation as it contained higher amounts of biodegradable matter. The calculated percentage of volatile solids loss for blend with newspaper was only 9.98% compared to 28.3% for blend with onion peels. The volatile solids of both blends were significantly different at $p = 0.0004$, with onion peels having lower means compared to the newspaper. Figure 3b presents that the percentage of volatile solids loss for the 2 kg was higher than that of the 6 kg, at 24.5% and 21.4%, respectively. These findings were lower than those of Cekmecelioglu et al. (2005) who discovered that the percent volatile solids loss of food waste in various windrow systems was in the range of 32.6% to 52.6%. The means of volatile solids content of the two load sizes were significantly different at $p = 0.0427$, where the 2-kg load size contained lower volatile solids than the 6 kg. With the same volatile solids contents of 87%, the compost with added starter culture attained a higher volatile solids loss of 22.5% compared to the control at 21.7% (Figure 3c). The starter culture could have been involved in degrading the materials slightly faster, although the average values between the two types of composts were not significantly different at $p = 0.4511$. This is because the starter culture serves as an inoculum, increasing bacterial density and perhaps reducing lag phase duration.

7.3.3 COMPOST MATURITY EVALUATION

The CN ratio is an accurate indication of compost maturity (Gray et al. 1971; Chanyasak and Kubota 1981; Jimenez and Garcia 1989). In these kitchen-waste composting studies, the TOC content in the compost material reduced arbitrarily higher than the reduction of total nitrogen contents resulting in a corresponding reduction of CN ratio as composting proceeds. The carbon provides the primary energy source for microbial metabolism, and nitrogen is critical for microbial population growth (Stoffella and Kahn 2001).

Figure 4a shows that the TOC content of the onion peel blend decreased to 28%, while the newspaper blend was at 10% over the 30-day period. The reduction of TOC is mainly because of the mineralisation process (Grigatti et al. 2004), a process where microbes employ organic matter and leave behind inorganic substances such as minerals, carbon dioxide and water. The blend with onion peels was prone to a higher mineralisation than the blend with newspaper because newspaper contains cellulose fibres sheathed in lignin, which is a compound found in wood with highly resistant to biological degradation (Trautmann et al. 1996). In addition, the thermophilic bacteria ($p < 0.0001$) and the thermophilic fungi ($p = 0.0002$) were governing the blend with onion peels, as they help accelerate the composting. In the blend with newspaper, the mesophilic fungi ($p < 0.0001$) was dominating the 30-day period of composting, which indicates that the mineralisation is low and takes a longer time for composting. The nitrogen content for the newspaper blend fluctuated at the range of 0.8% to 1.3%, while the onion peels increased to 4.2%. Both average carbon and nitrogen contents were significantly different with $p = 0.0004$ and $p < 0.0001$, respectively, with DMRT results showing lower carbon content and higher nitrogen content for the blend with onion peels. The resulting CN ratio in Figure 4c shows that the blend with newspaper underwent a slow composting process with its CN ratio value reduced to 37.04 compared to the blend with onion peels with a CN ratio of 8.15 after 30 days. Benito et al. (2006) stated that a CN ratio between 14 and 24 would be ideal for ready-to-use compost. For these reasons, the final products compost with onion peels were ready to be used on day 8 due to its CN ratio of 16.2, but the blend with newspaper needed more time for decomposition activity. The means of CN ratio for both blends were significantly different ($p < 0.0001$), with the onion peel blend having lower means of CN ratio than the newspaper. The initial increase in CN ratio of the newspaper blend could be due to the incomplete degradation process, and it did not reach thermophilic temperatures.

For load size studies, the 2-kg compost had lower TOC content and higher nitrogen content than the 6 kg (Figure 5a,b). The CN ratio of the 2-kg load increased slightly after 24 h before dropping rapidly to 10.94 on day 10 as compared to the 6 kg which required 30 days to reach a CN ratio of 10.15 (Figure 5c). The 2-kg load size decomposed more quickly compared to the 6-kg load size. The final CN ratio of the 2- and 6-kg loads was 9.22 and 10.15, respectively. They were considered matured and suitable to be applied to soil based on the study of Shiralipour et al. (1992) who have identified that if the CN ratio is less than 30, it will help crop production with its microbial mobilisation. Molnar and Bartha (1988) believed that a stable and high quality compost has a CN ratio in the range of 15 to 17. Hence, it was suggested that the end product of the 2-kg load size can be collected on day 8, half the time of the 6-kg load size on day 16. The means of the TOC content for the 2- and 6-kg load sizes were significantly different with $p = 0.0427$. The means of the total nitrogen content and CN ratio between the 2- and 6-kg load sizes, however, were not significantly different at $p = 0.0533$ and $p = 0.0701$, respectively.

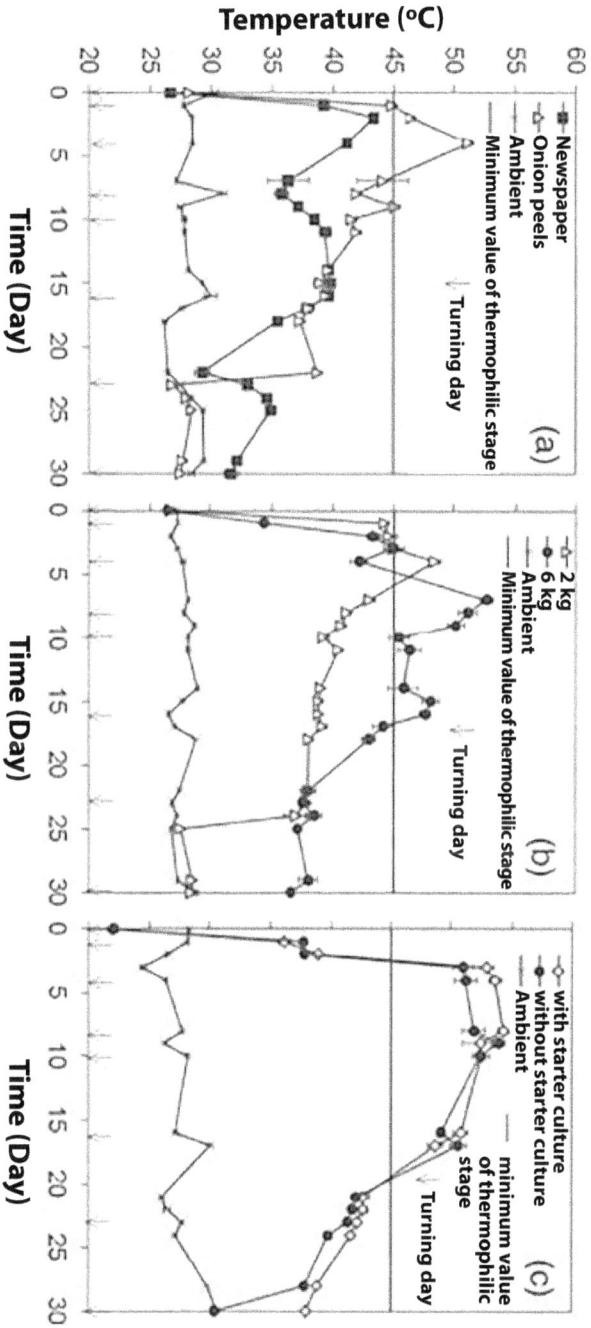

FIGURE 2: Temperature profiles of the compost. (a) With blends of newspaper or onion peels at 2-kg load size, (b) at 2- or 6-kg load size both with onion peels, and (c) with or without starter culture both at 6-kg load size using onion peels.

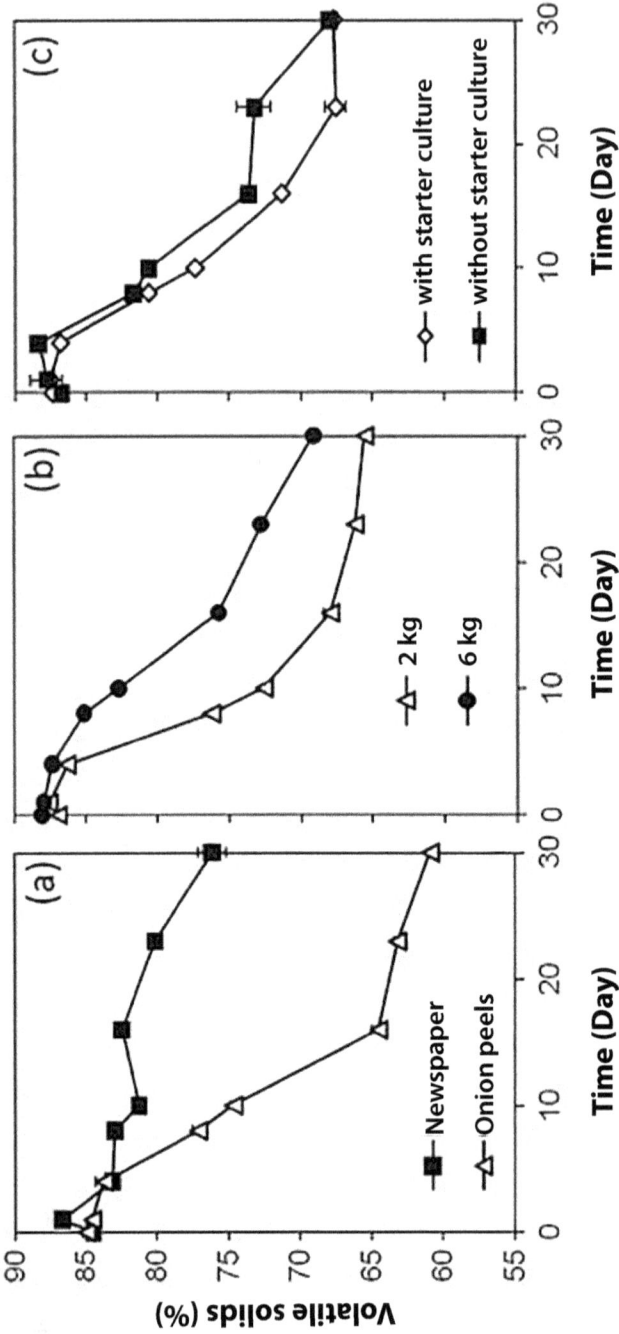

FIGURE 3: Volatile solids of compost. (a) With blends of newspaper or onion peels at 2-kg load size, (b) at 2- or 6-kg load size both with onion peels, and (c) with or without starter culture both at 6-kg load size using onion peels.

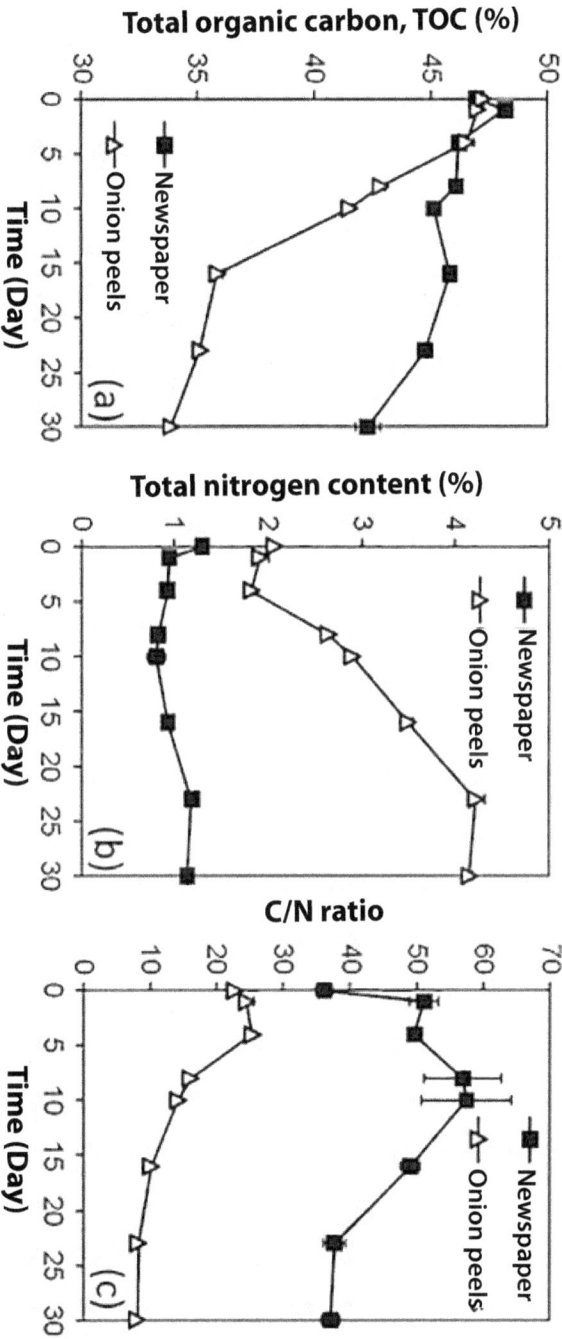

FIGURE 4: Composting using different blends of bulking media, newspaper or onion peels at 2-kg load size. (a) TOC, (b) nitrogen content and (c) CN ratio profiles.

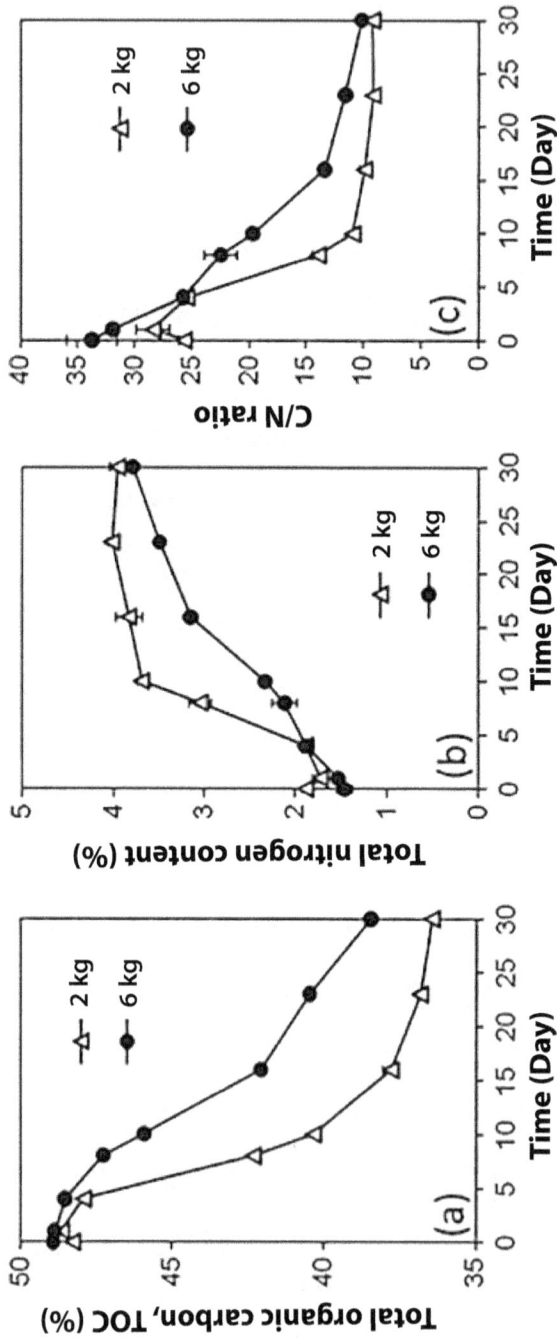

FIGURE 5: Composting using different load size of 2 or 6 kg with onion peels. (a) TOC, (b) nitrogen content and (c) CN ratio profiles.

Figure 6 shows that the mixture with added starter culture and control have no significant difference in terms of TOC content (p = 0.4512), total nitrogen content (p = 0.9183) and CN ratio (p = 0.8158). The starter culture has just a slight influence on the reduction of TOC due to the presence of microbial growth, although thermophilic temperatures do increase the level of carbonaceous material (Ginnivan et al. 1981). The microbial growth caused a carbon substrate limited condition even though the thermophilic temperatures could increase the level of carbon. The compost with added starter culture had lower nitrogen contents before day 10 and higher nitrogen contents after day 10 when compared to the control. The presence of microbes could have accelerated the decomposition activities in the beginning of the composting process. At the early composting stage, the mesophilic bacteria dominated the control up to 9 log colony-forming unit (CFU)/g, while the thermophilic bacteria dominated the mixture with added starter culture up to 13 log CFU/g. Both the TOC and nitrogen contents of the compost with added culture and control reached similar values at the end of composting. The decrease in CN ratio generally is explained by the bigger loss of TOC to produce carbon dioxide and smaller increase of nitrogen content as the decomposition progressed. The decreasing pattern of the CN ratio and TOC, and the increasing pattern of nitrogen content were similar to the findings of Goyal et al. (2005) on sugarcane trash, cattle dung, press mud, poultry waste and water hyacinth composting. The slight nitrogen losses of the control on day 10 which caused the slight CN ratio increase was due to nitrogen losses in the form of ammonia. This interpretation is consistent with Goyal et al. (2005) who found that nitrogen losses during ammonia volatilisation caused the increase of the CN ratio. The compost with added starter culture and control both possessed quite similar CN ratio of 9.33 and 9.35, respectively, which are less than 12 and in the range of acceptable degree of maturation (Bernal et al. 1998; Paredes et al. 2005). The optimum CN ratio of 10 has been suggested by Pöpel and Ohnmacht (1972) for the complete oxidation of the waste by aerobic bacteria.

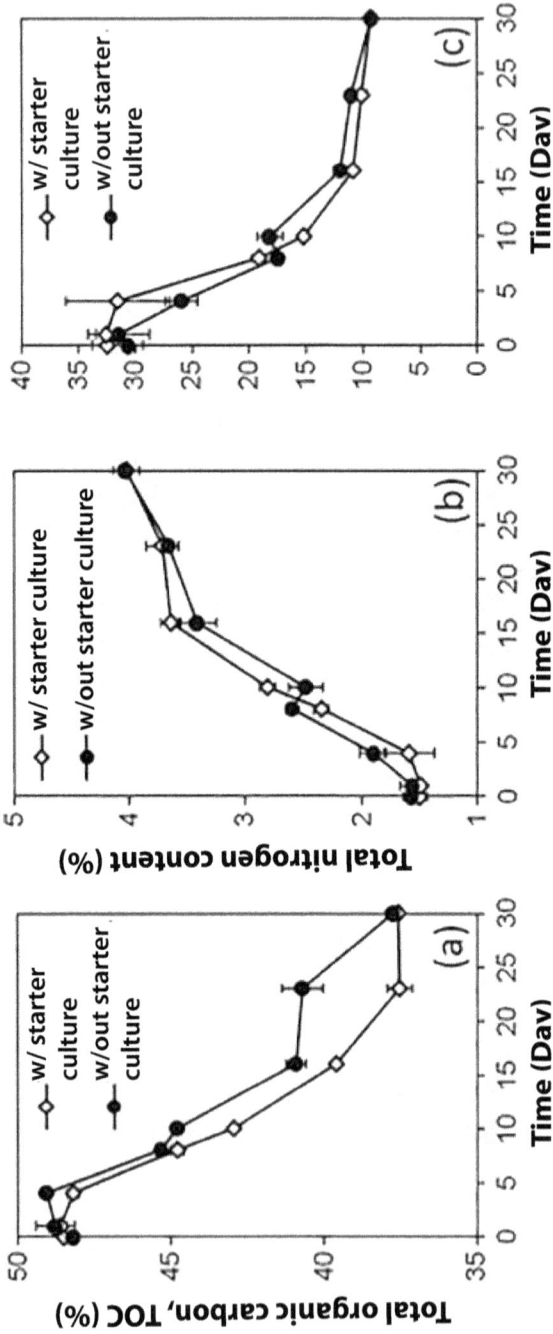

FIGURE 6: Composting using microbial treatments at 6-kg load size using onion peels. (a) TOC, (b) nitrogen content and (c) CN ratio profiles.

7.4 CONCLUSIONS

The onion peels were more suitable; at a smaller waste load, compost maturity with CN ratio below 10 was attained more quickly. The larger compost load size plays a vital role to achieve higher temperature. However, the lack of oxygen may have influenced the loss of volatile solids, and the lack of ventilation certainly influenced the loss of nitrogen by ammonia volatilisation when the maximum volume for the reactor, which is 6-kg load size, was used. No apparent differences were found in compost with added starter culture. The kitchen-waste composting process is said to require no additional microbes as accelerant, although it did help to achieve a slightly higher thermophilic temperature during the early stage of composting.

REFERENCES

1. Abdullah N, Chin NL (2010) Simplex-centroid mixture formulation for optimised composting of kitchen waste. Bioresour Technol 101(21):8205-8210
2. Belyaeva ON, Haynes RJ (2009) Chemical, microbial and physical properties of manufactured soils produced by co-composting municipal green waste with coal fly ash. Bioresour Technol 100(21):5203-5209
3. Benito M, Masaguer A, Moliner A, De Antonio R (2006) Chemical and physical properties of pruning waste compost and their seasonal variability. Bioresour Technol 97:2071-2076
4. Bernal PM, Navarro AF, Sánchez-Monedero MA, Roig A, Cegarra J (1998) Influence of sewage sludge compost stability and maturity on carbon and nitrogen mineralization in soil. Soil Biol Biochem 30(3):305-313
5. Bhatt R, Kausadikar H (2010) Definition of soil microbiology and soil in view of microbiology. http://www.scribd.com/doc/27429437/Soil-Microbiology . Accessed 18 Feb 2012
6. Cayuela ML, Sánchez-Monedero MA, Roig A (2006) Evaluation of two different aeration systems for composting two-phase olive mill wastes. Process Biochem 41(3):616-623
7. Cekmecelioglu D, Demirci A, Graves RE, Davitt NH (2005) Applicability of optimised in-vessel food waste composting for windrow systems. Biosyst Eng 91(4):479-486
8. Chang JI, Hsu TE (2008) Effects of compositions on food waste composting. Bioresour Technol 99(17):8068-8074
9. Chanyasak V, Kubota H (1981) Carbon/organic nitrogen ratio in water extract as measure of compost degradation. J Ferment Technol 59(3):215-219

10. Chikae M, Ikeda R, Kerman K, Morita Y, Tamiya E (2006) Estimation of maturity of compost from food wastes and agro-residues by multiple regression analysis. Bioresour Technol 97(16):1979-1985

11. Fang M, Wong MH, Wong JWC (2001) Digestion activity of thermophilic bacteria isolated from ash-amended sewage sludge compost. Water Air Soil Pollut 126(1–2):1-12

12. Ginnivan MJ, Woods JL, O'Callaghan JR (1981) Thermophilic aerobic treatment of pig slurry. J Agric Eng Res 26(6):455-466

13. Goyal S, Dhull SK, Kapoor KK (2005) Chemical and biological changes during composting of different organic wastes and assessment of compost maturity. Bioresour Technol 96(14):1584-1591

14. Gray KR, Sherman K, Biddlestone AJ (1971) A review of composting. Part I. Microbiology and biochemistry. Process Biochem 6:32-36

15. Grigatti M, Ciavatta C, Gessa C (2004) Evolution of organic matter from sewage sludge and garden trimming during composting. Bioresour Technol 91(2):163-169

16. Huang DL, Zeng GM, Feng CL, Hu S, Lai C, Zhao MH, Su FF, Tang L, Liu HL (2010) Changes of microbial population structure related to lignin degradation during lignocellulosic waste composting. Bioresour Technol 101(11):4062-4067

17. Ishii K, Takii S (2003) Comparison of microbial communities in four different composting processes as evaluated by denaturing gradient gel electrophoresis analysis. J Appl Microb 95:109-119

18. Jimenez EI, Garcia VP (1989) Evaluation of city refuse compost maturity: a review. Biol Wastes 27:115-142

19. Kalamdhad AS, Pasha M, Kazmi AA (2008) Stability evaluation of compost by respiration techniques in a rotary drum composter. Res Cons Recycl 52(5):829-834

20. Kamil NK, Alwi SA, Singh M (2010) Malaysia. Malaysia. http://www.readbag.com/avrdc-pdf-dynamics-malaysia . Accessed 18 Feb 2012

21. Khan MAI, Ueno K, Horimoto S, Komai F, Tanaka K, Ono Y (2009) Physicochemical, including spectroscopic, and biological analyses during composting of green tea waste and rice bran. Biol Fert Soils 45(3):305-313

22. Laos F, Mazzarino MJ, Walter I, Roselli L, Satti P, Moyano S (2002) Composting of fish offal and biosolids in northwestern Patagonia. Bioresour Technol 81(3):179-186

23. Lubberding HJ, Gijzen HJ, Heck M, Vogels GD (1988) Anaerobic digestion of onion waste by means of rumen microorganisms. Biol Wastes 25(1):61-67

24. Luostarinen S, Rintala J (2007) Anaerobic on-site treatment of kitchen waste in combination with black water in UASB-septic tanks at low temperatures. Bioresour Technol 98:1734-1740

25. Mohee R, Driver M-FB, Sobratee N (2008) Transformation of spent broiler litter from exogenous matter to compost in a sub-tropical context. Bioresour Technol 99(1):128-136

26. Molnar L, Bartha I (1988) High solids anaerobic fermentation for biogas and compost production. Biomass 16(3):173-182

27. Naik VN, Sharma DD, Kumar PMP, Yadav RD (2012) Efficacy of ligno-cellulolytic fungi on recycling sericultural wastes. Acta Biologica Indica 1(1):47-50

28. Pagans E, Barrena R, Font X, Sánchez A (2006) Ammonia emissions from the composting of different organic wastes. Dependency on process temperature. Chemosphere 62(9):1534-1542

29. Palmisano AC, Maruscik DA, Ritchie CJ, Schwab BS, Harper SR, Rapoport RA (1993) A novel bioreactor simulating composting of municipal solid waste. J Microbiol Meth 18(2):99-112

30. Paredes C, Cegarra J, Bernal MP, Roig A (2005) Influence of olive mill wastewater in composting and impact of the compost on a Swiss chard crop and soil properties. Environ Int 31(2):305-312

31. Pöpel F, Ohnmacht C (1972) Thermophilic bacterial oxidation of highly concentrated substrates. Water Res 6(7):807-815

32. Raut MP, Prince William SPM, Bhattacharyya JK, Chakrabarti T, Devotta S (2008) Microbial dynamics and enzyme activities during rapid composting of municipal solid waste—a compost maturity analysis perspective. Bioresour Technol 99(14):6512-6519

33. Rogoshewski P, Bryson H, Wagner K (1983) Remedial action technology for waste disposal sites. Park Ridge, New Jersey: Noyes Data Corporation.

34. Sellami F, Jarboui R, Hachicha S, Medhioub K, Ammar E (2008) Co-composting of oil exhausted olive-cake, poultry manure and industrial residues of agro-food activity for soil amendment. Bioresour Technol 99(5):1177-1188

35. Shiralipour A, McConnell DB, Smith WH (1992) Physical and chemical properties of soils as affected by municipal solid waste compost application. Biomass Bioenergy 3(3–4):261-266

36. Stabnikova O, Ding H-B, Tay J-H, Wang J-Y (2005) Biotechnology for aerobic conversion of food waste into organic fertilizer. Waste Manage Res 23:39-47

37. Stoffella PJ, Kahn BA (2001) Compost utilization in horticultural cropping systems. Boca Raton: Lewis Publishers.

38. Sun Y, Cheng J (2002) Hydrolysis of lignocellulosic materials for ethanol production: a review. Bioresour Technol 83(1):1-11

39. Sundberg C, Jönsson H (2005) Process inhibition due to organic acids in fed-batch composting of food waste—influence of starting culture. Biodegradation 16:205-213

40. Sundberg C, Smårs S, Jönsson H (2004) Low pH as an inhibiting factor in the transition from mesophilic to thermophilic phase in composting. Bioresour Technol 95(2):145-150

41. Suutarinen M, Mustranta A, Autio K, Salmenkallio-Marttila M, Ahvenainen R, Buchert J (2003) The potential of enzymatic peeling of vegetables. J Sci Food Agric 83(15):1556-1564

42. Trautmann NM, Richard T, Krasny ME (1996) Cornell composting: science and engineering. Compost chemistry. http://compost.css.cornell.edu/chemistry.html . Accessed 18 Feb 2012

43. Unmar G, Mohee R (2008) Assessing the effect of biodegradable and degradable plastics on the composting of green wastes and compost quality. Bioresour Technol 99(15):6738-6744

44. Veeken A, Hamelers B (1999) Effect of temperature on hydrolysis rates of selected biowaste components. Bioresour Technol 69:249-254

45. Volchatova IV, Belovezhets LA, Medvedeva SA (2002) Microbiological and biochemical investigation of succession in lignin-containing compost piles. Microbiology 71(4):467-470

46. Wang JY, Stabnikova O, Ivanov V, Tay STL, Tay JH (2003) Intensive aerobic bioconversion of sewage sludge and food waste into fertiliser. Waste Manage Res 21(5):405-415

47. Wang Q, Yamabe K, Narita J, Morishita M, Ohsumi Y, Kusano K, Shirai Y, Ogawa HI (2001) Suppression of growth of putrefactive and food poisoning bacteria by lactic acid fermentation of kitchen waste. Process Biochem 37:351-357

48. Wayman M, Chen S, Doan K (1992) Bioconversion of waste paper to ethanol. Process Biochem 27(4):239-245

CHAPTER 8

Microbial Diversity of Vermicompost Bacteria that Exhibit Useful Agricultural Traits and Waste Management Potential

JAYAKUMAR PATHMA AND NATARAJAN SAKTHIVEL

8.1 INTRODUCTION

Soil, is the soul of infinite life that promotes diverse microflora. Soil bacteria viz., *Bacillus*, *Pseudomonas* and *Streptomyces* etc., are prolific producers of secondary metabolites which act against numerous co-existing phytopathogeic fungi and human pathogenic bacteria (Pathma et al. 2011b). Earthworms are popularly known as the "farmer's friend" or "nature's plowman". Earthworm influences microbial community, physical and chemical properties of soil. They breakdown large soil particles and leaf litter and thereby increase the availability of organic matter for microbial degradation and transforms organic wastes into valuable vermicomposts by grinding and digesting them with the help of aerobic and anaerobic microbes (Maboeta and Van Rens-

Microbial Diversity of Vermicompost Bacteria that Exhibit Useful Agricultural Traits and Waste Management Potential. © *Pathma J and Sakthivel N.* SpringerPlus *1,26 (2012), doi:10.1186/2193-1801-1-26. Licensed under Creative Commons Attribution 2.0 Generic License, http://creativecommons.org/licenses/by/2.0/.*

burg 2003). Earthworms activity is found to enhance the beneficial micro-flora and suppress harmful pathogenic microbes. Soil wormcasts are rich source of micro and macro-nutrients, and microbial enzymes (Lavelle and Martin 1992). Vermicomposting is an efficient nutrient recycling process that involves harnessing earthworms as versatile natural bioreactors for organic matter decomposition. Due to richness in nutrient availability and microbial activity vermicomposts increase soil fertility, enhance plant growth and suppress the population of plant pathogens and pests. This review paper describes the bacterial biodiversity and nutrient status of vermicomposts and their importance in agriculture and waste management.

8.1.1 EARTHWORMS

Earthworms are capable of transforming garbage into 'gold'. Charles Darwin described earthworms as the 'unheralded soldiers of mankind', and Aristotle called them as the 'intestine of earth', as they could digest a wide variety of organic materials (Darwin and Seward 1903; Martin 1976). Soil volume, microflora and fauna influenced by earthworms have been termed as "drilosphere" and the soil volume includes the external structures produced by earthworms such as surface and below ground casts, burrows, middens, diapause chambers as well as the earthworms body surface and internal gut associated structures in contact with the soil (Lavelle et al. 1989; Brown et al. 2000). Earthworms play an essential role in carbon turnover, soil formation, participates in cellulose degradation and humus accumulation. Earthworm activity profoundly affects the physical, chemical and biological properties of soil. Earthworms are voracious feeders of organic wastes and they utilize only a small portion of these wastes for their growth and excrete a large proportion of wastes consumed in a half digested form (Edwards and Lofty 1977; Kale and Bano 1986; Jambhekar 1992). Earthworms intestine contains a wide range of microorganisms, enzymes and hormones which aid in rapid decomposition of half-digested material transforming them into vermicompost in a short time (neary 4–8 weeks) (Ghosh et al. 1999; Nagavallemma et al. 2004) compared to traditional composting process which takes the advantage of microbes alone and thereby requires a prolonged period (nearly 20 weeks) for compost production (Bernal et al. 1998; Sánchez-Monedero et al. 2001). As the organic matter passes through the gizzard of the earthworm it is grounded into a fine powder after which the digestive enzymes, microorganisms and other fermenting substances act on them further aiding their breakdown within the gut, and finally passes out in the form of "casts" which are later acted upon by earthworm gut associated microbes converting them into mature product, the "vermicomposts" (Dominguez and Edwards 2004).

Earthworms, grouped under phylum annelida are long, narrow, cylindrical, bilaterally symmetrical, segmented soil dwelling invertebrates with a glistening dark brown body covered with delicate cuticle. They are hermaphrodites and weigh over 1,400–1,500 mg after 8–10 weeks. Their body contains 65% protein (70–80% high quality 'lysine rich protein' on a dry weight basis), 14% fats, 14% carbohydrates, and 3% ash. Their life span varies between 3–7 years depending upon the species and ecological situation. The gut of earthworm is a straight tube starting from mouth followed by a muscular pharynx, oesophagus, thin walled crop, muscular gizzard, foregut, midgut, hindgut, associated digestive glands, and ending with anus. The gut consisted of mucus containing protein and polysaccharides, organic and mineral matter, amino acids and microbial symbionts viz., bacteria, protozoa and microfungi. The increased organic carbon, total organic carbon and nitrogen and moisture content in the earthworm gut provide an optimal environment for the activation of dormant microbes and germination of endospores etc. A wide array of digestive enzymes such as amylase, cellulase, protease, lipase, chitinase and urease were reported from earthworm's alimentary canal. The gut microbes were found to be responsible for the cellulase and mannose activities (Munnoli et al. 2010). Earthworms comminutes the substrate, thereby increases the surface area for microbial degradation constituting to the active phase of vermicomposting. As this crushed organic matter passes through the gut it get mixed up with the gut associated microbes and the digestive enzymes and finally leaves the gut in partially digested form as "casts" after which the microbes takes up the process of decomposition contributing to the maturation phase (Lazcano et al. 2008).

Association of earthworms with microbes is found to be complex. Certain groups of microbes were found to be a part of earthworm's diet which is evidenced by the destruction of certain microbes as they pass through the earthworms digestive system. Few yeasts, protozoa and certain groups of fungi such as *Fusarium oxysporum, Alternaria solani,* and microfungi were digested by the earthworms, *Drawida calebi, Lumbricus terrestris* and *Eisenia foetida. Bacillus cereus* var *mycoides* were reported to decrease during gut passage while *Escherichia coli* and *Serratia marcessens* were completely eliminated during passage through earthworm gut (Edwards and Fletcher, 1988).

TABLE 1: Ecological categories and niches of earthworms and their characteristic features and beneficial traits

Species	Ecological category	Ecological niche	Characteristic features	Beneficial trait
Eisenia foetida,	Epigeics	Superficial soil layers, leaf litter, compost	Smaller in size, body uniformly pigmented, active gizzard, short life cycle, high reproduction rate and regeneration, tolerant to disturbance, phytophagous	Efficient bio-degraders and nutrient releasers, efficient compost producers, aids in litter comminution and early decomposition
Lumbricus rubellus,				
L. castaneus,				
L. festivus,				
Eiseniella tetraedra,				
Bimastus minusculus,				
B. eiseni,				
Dendrodrilus rubidus,				
Dendrobaena veneta,				
D. octaedra				
Aporrectodea caliginosa,	Endogeics	Topsoil or subsoil	Small to large sized worms, weakly pigmented, life cycle of medium duration, moderately tolerant to disturbance, geophagous	Brings about pronounced changes in soil physical structure, can efficiently utilize energy from poor soils hence can be used for soil improvements

TABLE 1: *Cont.*

Species	Ecological category	Ecological niche	Characteristic features	Beneficial trait
A. trapezoides,				
A. rosea,				
Millsonia anomala,				
Octolasion cyaneum,	Polyhumic endogeic	Top soil (A1)	Small size, unpigmented, forms horizontal burrows, rich soil feeder	
O. lacteum,				
Pontoscolex corethrurus,	Meso-humic endogeic	A and B horizon	Medium size, unpigmented, forms extensive horizontal burrows, bulk (A1) soil feeder	
Allolobophora chlorotica,				
Aminthas sp.	Oligo-humic endogeic	B and C horizon	Very large in size, unpigmented, forms extensive horizontal burrows, feeds on poor, deep soils	
L. terrestris,	Anecics	Permanent deep burrows in soil	Large in size, dorsally pigmented, forms extensive, deep, vertical permanent burrows, low reproductive rate, sensitive to disturbance, phytogeophagous, nocturnal	Forms vertical burrows affecting air-water relationship and movement from deep layers to surface helps in efficient mixing of nutrients
L. polyphemus,				
A. longa				

Earthworms are classified into epigeic, anecic and endogeic species based on definite ecological and trophic functions (Brown 1995; Bhatnagar and Palta 1996) (Table 1). Epigeic earthworms are smaller in size, with uniformly pigmented body, short life cycle, high reproduction rate and regeneration. They dwell in superficial soil surface within litters, feeds on the surface litter and mineralize them. They are phytophagous and rarely ingest soil. They contain an active gizzard which aids in rapid conversion of organic matter into vermicomposts. In addition epigeic earthworms are efficient bio-degraders and nutrient releasers, tolerant to disturbances, aids in litter comminution and early decomposition and hence can be efficiently used for vermicomposting. Epigeic earthworms includes *Eisenia foetida, Lumbricus rubellus, L. castaneus, L. festivus, Eiseniella tetraedra, Bimastus minusculus, B. eiseni, Dendrodrilus rubidus, Dendrobaena veneta, D. octaedra.* Endogeics earthworms are small to large sized worms, with weakly pigmented body, life cycle of medium duration, moderately tolerant to disturbance, forms extensive horizontal burrows and they are geophagous feeding on particulate organic matter and soil. They bring about pronounced changes in soil physical structure and can efficiently utilize energy from poor soils, hence can be used for soil improvements. Endogeics include *Aporrectodea caliginosa, A. trapezoides, A. rosea, Millsonia anomala, Octolasion cyaneum, O. lacteum, Pontoscolex corethrurus, Allolobophora chlorotica* and *Aminthas* sp. They are further classified into polyhumic endogeic which are small sized, rich soil feeding earthworms, dwelling in top soil (A1); mesohumic endogeic which are medium sized worms, dwelling in A and B horizon, feeding on bulk (A1) soil; and oligohumic endogeic which are very large worms, dwelling in B and C horizons, feeding on poor, deep soil. Aneceics are larger, dorsally pigmented worms, with low reproductive rate, sensitive to disturbance, nocturnal, phytogeophagous, bury the surface litter, forms middens and extensive, deep, permanent vertical burrows, and live in them. Formation of vertical burrows affects airwater relationship and movement from deep layers to surface helps in efficient mixing of nutrients. *Lumbricus terrestris, L. polyphemus* and *Aporrectodea longa* are examples of aneceics earthworms (Kooch and Jalilvand 2008). Epigeics and aneceics are harnessed largely for vermicomposting (Asha et al. 2008). Epigeics namely *Eisenia foetida* (Hartenstein et al. 1979), *Eudrilus eugeniae* (Kale and Bano 1988), *Perionyx excavatus* (Sinha et al. 2002; Suthar and Singh 2008) and *Eisenia anderi* (Munnoli et al. 2010) have been used in converting organic wastes into vermicompost. These surface dwellers capable of working on litter layers converting them into manure are of no significant value in modifying the soil structure. In contrast, anecics such as *Lampito mauritii* are efficient creators of an effective drilosphere as well as excellent compost producers (Ismail 1997). Earthworms thus act as natural bio-reactors, altering the nature of the organic waste by fragmenting them.

Earthworm activity engineers the soil by forming extensive burrows which loosen the soil and makes it porous. These pores improve aeration, water absorption, drainage and easy root penetration. Soil aggregates formed by earthworms and associated microbes, in the casts and burrow walls play an indispensible role in soil air ecosystem. These aggregates are mineral granules bonded in a way to resist erosion and to avoid soil compaction both in wet and dry condition. Earthworms speed up soil reclamation and make them productive by restoring beneficial microflora (Nakamura 1996). Thus degraded unproductive soils and land degraded by mining could be engineered physically, chemically and biologically and made productive by earthworms. Hence earthworms are termed as ecosystem engineers (Brown et al. 2000; Munnoli et al. 2010).

8.1.2 VERMICOMPOSTING

Vermicomposting is a non-thermophilic biological oxidation process in which organic material are converted into vermicompost which is a peat like material, exhibiting high porosity, aeration, drainage, water holding capacity and rich microbial activities (Edwards 1998; Atiyeh et al. 2000b; Arancon et al. 2004a), through the interactions between earthworms and associated microbes. Vermiculture is a cost-effective tool for environmentally sound waste management (Banu et al. 2001; Asha et al. 2008). Earthworms are the crucial drivers of the process, as they aerate, condition and fragment the substrate and thereby drastically alter the microbial activity and their biodegradation potential (Fracchia et al. 2006; Lazcano et al. 2008). Several enzymes, intestinal mucus and antibiotics in earthworm's intestinal tract play an important role in the breakdown of organic macromolecules. Biodegradable organic wastes such as crop residues, municipal, hospital and industrial wastes pose major problems in disposal and treatment. Release of unprocessed animal manures into agricultural fields contaminates ground water causing public health risk. Vermicomposting is the best alternative to conventional composting and differs from it in several ways (Gandhi et al. 1997). Vermicomposting hastens the decomposition process by 2–5 times, thereby quickens the conversion of wastes into valuable biofertilizer and produces much more homogenous materials compared to thermophilic composting (Bhatnagar and Palta 1996; Atiyeh et al. 2000a). Distinct differences exist between the microbial communities found in vermicomposts and composts and hence the nature of the microbial processes is quite different in vermicomposting and composting (Subler et al. 1998). The active phase of composting is the thermophilic stage characterized by thermophilic bacterial community where intensive decomposition takes place followed by a mesophilic maturation phase (Lazcano et al. 2008; Vivas et al. 2009). Vermicomposting is a mesophilic process characterized by mesophilic bacteria and fungi (Benitez et al. 1999). Vermicomposting comprises of an active stage during which earthworms and associated microbes jointly process the substrate and the maturation phase that involves the action of associated microbes and occurs once the worm's moves to the fresher layers of undigested waste or when the product is removed from the vermireacter. The duration of the active phase depends on the species and density of the earthworms involved (Ndegwa et al. 2000; Lazcano et al. 2008; Aira et al. 2011). A wide range of oganic wastes viz., horticultural residues from processed potatoes (Edwards 1988); mushroom wastes (Edwards 1988; Tajbakhsh et al. 2008); horse wastes (Hartenstein et al. 1979; Edwards et al. 1998); pig wastes (Chan and Griffiths 1988; Reeh 1992); brewery wastes (Butt 1993); sericulture wastes (Gunathilagraj and Ravignanam 1996); municipal sewage sludge (Mitchell et al. 1980; Dominguez et al. 2000); agricultural residues (Bansal and Kapoor 2000); weeds (Gajalakshmi et al. 2001); cattle dung (Gunadi et al. 2002); industrial refuse such as paper wastes (Butt 1993; Elvira et al. 1995; Gajalakshmi et al. 2002); sludge from paper mills and dairy plants (Elvira et al. 1997; Banu et al. 2001); domestic kitchen wastes (Sinha et al. 2002); urban residues and animal wastes (Edwards et al. 1985; Edwards 1988) can be vermicomposted (Sharma et al. 2005).

Effects of vermicomposting on pH, electrical conductivity (EC), C:N ratio and other nutrients have been documented. Earthworm activity reduced pH and C:N ratio in manure (Gandhi et al. 1997; Atiyeh et al. 2000b). Chemical analysis showed vermicompost had a lower pH, EC, organic carbon (OC) (Nardi et al. 1983; Albanell et al. 1988; Mitchell 1997),

C:N ratio (Riffaldi and Levi-Minzi 1983; Albanell et al. 1988), nitrogen and potassium and higher amounts of total phosphorous and micronutrients compared to the parent material (Hashemimajd et al. 2004). Slightly decreased pH values of vermicompost compared to traditional compost might be attributed due to mineralization of N and P, microbial decomposition of organic materials into intermediate organic acids, fulvic acids, humic acids (Lazcano et al. 2008; Albanell et al. 1988; Chan and Griffiths 1988; Subler et al. 1998) and concomitant production of CO_2 (Elvira et al. 1998; Garg et al. 2006). Vermicomposting of paper mill and dairy sludge resulted in 1.2–1.7 fold loss of organic carbon as CO_2 (Elvira et al. 1998). In contrast to the parent material used, vermicomposts contain higher humic acid substances (Albanell et al. 1988). Humic acid substances occur naturally in mature animal manure, sewage sludge or paper-mill sludge, but vermicomposting drastically increases the rate of production and their amount from 40–60 percent compared to traditional composting. The enhancement in humification processes is by fragmentation and size reduction of organic matter, increased microbial activity within earthworm intestine and soil aeration by earthworm feeding and movement (Dominguez and Edwards, 2004). EC indicates the salinity of the organic amendment. Minor production of soluble metabolites such as ammonium and precipitation of dissolved salts during vermicomposting lead to lower EC values. Compared to the parent material used, vermicomposts contain less soluble salts and greater cation exchange capacity (Holtzclaw and Sposito 1979; Albanell et al. 1988). C:N ratio is an indicator of the degree of decomposition. During the process of biooxidation, CO_2 and N is lost and loss of N takes place at a comparatively lower rate. Comparison of compost and vermicompost showed that vermicompost had significantly less C:N ratios as they underwent intense decomposition (Lazcano et al. 2008).

Vermicomposting of cow manure using earthworm species *E. andrei* (Atiyeh et al. 2000b) and *E. foetida* (Hand et al. 1988) favored nitrification, resulting in the rapid conversion of ammonium-nitrogen to nitrate-nitrogen. Vermicomposting increased the concentration of nitrate-nitrogen to 28 fold after 17 weeks, while in conventional compost there was only 3-fold increase (Subler et al. 1998; Atiyeh et al. 2000a). Increase in ash concentration during vermicomposting suggests that vermicomposting accelerates the rate of mineralization (Albanell et al. 1988). Mineralization is the process in which the chemical compounds in the organic matter decompose or oxidise into forms that could be easily assimilated by the plants. Increase in ash content increases the rate of mineralization. Ash is an alkaline substance which hinders the formation of H_2S as well as improves the availibility of O_2 and thereby renders composts odorless. Thus vermicomposting increases the ash content and accelerates the rate of mineralization which is essential to make nutrients available to plants. The observed increase of total phosphorous (TP) in vermicompost is probably due to mineralization and mobilization of phosphorus resulting from the enhanced phosphatase activity by microorganisms in the gut epithelium of the earthworms (Zhang et al. 2000; Garg et al. 2006). Vermicomposts showed a significant increase in exchangeable Ca^{2+}, Mg^{2+} and K^+ compared to fresh sludge indicating the conversion of nutrients to plant-available forms during passage through the earthworm gut (Garg et al. 2006; Yasir et al. 2009a). Vermicomposts contain higher nutrient concentrations, but less likely to produce salinity, than composts. Additionally, vermicomposts possess outstanding biological properties and have microbial populations significantly larger and more diverse compared to conventional composts (Edwards 1998). Soil supplemented with vermicompost showed better plant growth compared to soil treated with inorganic fertilizers or cattle manure (Kalembasa 1996; Subler et al. 1998).

8.1.3 DIVERSITY OF BACTERIA ASSOCIATED WITH EARTHWORMS

Earthworm's ability to increase plant nutrient availability is likely to be dependent on the activity of earthworm gut microflora. Earthworms indirectly influence the dynamics of soil chemical processes, by comminuting the litter and affecting the activity of the soil micro-flora (Petersen and Luxton 1982; Lee 1985; Edwards and Bohlen 1996). Interactions between earthworms and microorganisms seem to be complex. Earthworms ingest plant growth-promoting rhizospheric bacteria such as *Pseudomonas, Rhizobium, Bacillus, Azosprillium, Azotobacter,* etc. along with rhizospheric soil, and they might get activated or increased due to the ideal micro-environment of the gut. Therefore earthworm activity increases the population of plant growth-promoting rhizobacteria (PGPR) (Sinha et al. 2010). This specific group of bacteria stimulates plant growth directly by solubilization of nutrients (Ayyadurai et al. 2007; Ravindra et al. 2008), production of growth hormone, 1-aminocyclopropane-1-carboxylate (ACC) deaminase (Correa et al. 2004), nitrogen fixation (Han et al. 2005), and indirectly by suppressing fungal pathogens. Antibiotics, fluorescent pigments, siderophores and fungal cell-wall degrading enzymes namely chitinases and glucanases (Han et al. 2005; Sunish et al. 2005; Ravindra et al. 2008; Jha et al. 2009; Pathma et al. 2010; Pathma et al. 2011a, b) produced by bacteria mediate the fungal growth-suppression. Earthworms are reported to have association with such free living soil bacteria and constitute the drilosphere (Ismail 1995). Earthworm microbes mineralize the organic matter and also facilitate the chelation of metal ions (Pizl and Novokova 1993; Canellas et al. 2002). Gut of earthworms *L. terrestris, Allolobophora caliginosa* and *Allolobophora terrestris* were reported to contain higher number of aerobes compared to soil (Parle 1963). Earthworms increased the number of microorganisms in soil as much as five times (Edwards and Lofty 1977) and the number of bacteria and 'actinomycetes' contained in the ingested material increased upto 1,000 fold while passing through their gut (Edwards and Fletcher 1988). Similar increase was observed in plate counts of total bacteria, proteolytic bacteria and actinomycetes by passage through earthworms gut (Parle 1963; Daniel and Anderson 1992; Pedersen and Hendriksen 1993; Devliegher and Verstraete 1995). Similarly microbial biomass either decreased (Bohlen and Edwards 1995; Devliegher and Verstraete 1995), or increased (Scheu 1992) or remained unchanged (Daniel and Anderson 1992) after passage through the earthworm gut. An oxalate-degrading bacterium *Pseudomonas oxalaticus* was isolated from intestine of *Pheretima* species (Khambata and Bhat 1953) and an actinomycete *Streptomyces lipmanii* was identified in the gut of *Eisenia lucens* (Contreras 1980). Scanning electron micrographs provided evidence for endogenous microflora in guts of earthworms, *L. terrestris* and *Octolasion cyaneum* (Jolly et al. 1993). Gut of *E. foetida* contained various anaerobic N_2-fixing bacteria such as *Clostridium butyricum, C. beijerinckii* and *C. paraputrificum* (Citernesi et al. 1977). Alimentary canal of *Lumbricus rubellus* and *Octolasium lacteum* were found to contain more numbers of aerobes and anerobes (Karsten and Drake 1995) and culturable denitrifiers

(Karsten and Drake 1997). List of vermicompost bacteria and their beneficial traits is presented in Table 2.

TABLE 2: Biodiversity of vermicompost bacteria and their beneficial traits

Vermicompost earthworm	Names of bacteria	Beneficial traits	References
Pheretima sp.	*Pseudomonas oxalaticus*	Oxalate degradation	Khambata and Bhat, 1953
Unspecified	*Rhizobium trifolii*	Nitrogen fixation and growth of leguminous plants	Buckalew et al. 1982
Lumbricus rubellus	*R. japonicum, P. putida*	Plant growth promotion	Madsen and Alexander 1982
L. terrestris	*Bradyrhizobium japonicum*	Improved distribution of nodules on soybean roots	Rouelle, 1983
Aporrectodea trapezoids,	*P. corrugata* 214OR	Suppress *Gaeuman-nomyces* graminis var. Tritd in wheat	Doube et al. 1994
A. rosea			
A. trapezoids,	*R. meliloti* L5-30R	Increased root nodulation and nitrogen fixation in legumes	Stephens et al. 1994b
Microscolex dubius			
Eisenia foetida	*Bacillus* spp., *B. megaterium, B. pumilus, B. subtilis*	Antimicrobial activity against Enterococcus faecalisDSM 2570, Staphylococcus aureus DSM 1104	Vaz-Moreira et al. 2008
L. terrestris	Fluorescent pseudomonads,	Suppress Fusarium oxysporum f. sp. asparagi and F. proliferatum in asparagus, Verticillium dahlia in eggplant and *F. oxysporum* f. sp. lycopersici Race 1 in tomato	Elmer, 2009
Filamentous actinomycetes			

TABLE 2: *Cont.*

Vermicompost earthworm	Names of bacteria	Beneficial traits	References
Eudrilus sp.	Free-living N₂ fixers, *Azospirillum, Azotobacter, Autotrophic Nitrosomonas, Nitrobacter,* Ammonifying bacteria, Phosphate solubilizers, Fluorescent pseudomonads	Plant growth promotion by nitrification, phosphate solubilisation and plant disease suppression	Gopal et al. 2009
E. foetida	*Proteobacteria, Bacteroidetes,*	Antifungal activity against Colletotrichum coccodes, *R. solani, P. ultimum, P. capsici* and *F. moliniforme*	Yasir et al. 2009a
	Verrucomicrobia, Actinobacteria,		
	Firmicutes		
Unspecified	*Eiseniicola composti* YC06271T	Antagonistic activity against *F. moniliforme*	Yasir et al. 2009b

Earthworms harbor 'nitrogen-fixing' and 'decomposer microbes' in their gut and excrete them along with nutrients in their excreta (Singleton et al. 2003). Earthworms stimulate and accelerate microbial activities by increasing the population of soil microorganisms (Binet et al. 1998), microbial numbers and biomass (Edwards and Bohlen 1996), by improving aeration through burrowing actions. Vermicomposting modified the original microbial community of the waste in a diverse way. Actinobacteria and Gammaproteobacteria were abundant in vermicompost, while conventional compost contained more Alphaproteobacteria and Bacteroidetes, the bacterial phylogenetic groups typical of non-cured compost (Vivas et al. 2009). Total bacterial counts exceeded 10^{-10}/ g of vermicompost and it included nitrobacter, azotobacter, rhizobium, phosphate solubilizers and actinomycetes (Suhane 2007). Molecular and culture-dependent analyses of bacterial community of vermicompost showed the presence of α-Proteobacteria, β-Proteobacteria, γ-Proteobacteria, Actinobacteria, Planctomycetes, Firmicutes and Bacteroidetes (Yasir et al. 2009a). Several findings showed considerable increase in total viable counts of actinomycetes and bacteria in the worm treated compost (Parthasarathi and Ranganathan 1998; Haritha Devi et al. 2009). The increase of microbial population may be due to the congenial condition for the growth of microbes within the digestive tract of earthworm and by the ingestion of nutrient rich organic wastes which provide energy and also act as a substrate for the growth of microorganisms (Tiwari et al. 1989). The differences in microbial species, numbers and activity between the earthworm alimentary canal or burrow and bulk soil indirectly support the hypothesis that the bacterial community structures of these habitats are different from those of the soil. Specific phylogenetic groups of bacteria such as *Aeromonas hydrophila* in *E. foetida* (Toyota and Kimura 2000), fluorescent pseudomonads in *L. terrestris* (Devliegher and Verstraete 1997), and Actinobacteria in *L. rubellus* (Kristufek et al. 1993) have been found in higher numbers in earthworm guts, casts, or burrows.

Enzymatic activity characterization and quantification has a direct correlation with type and population of microbes and reflects the dynamics of the composting process in terms of the decomposition of organic matter and nitrogen transformations and provide information about the maturity of the compost (Tiquia 2005). Wormcasts contain higher activities of cellulase, amylase, invertase, protease, peroxidase, urease, phosphatase and dehydrogenase (Sharpley and Syers 1976; Edwards and Bohlen 1996). Dehydrogenase is an intracellular enzyme related to the oxidative phosphorylation process (Trevors 1984) and is an indicator of microbial activity in soil and other biological ecosystems (Garcia et al. 1997). The maximum enzyme activities (cellulase, amylase, invertase, protease and urease) were observed during 21–35 days in vermicomposting and on 42–49 days in conventional composting. Also, microbial numbers and their extracellular enzyme profiles were more abundant in vermicompost produced from fruitpulp, vegetable waste, groundnut husk and cowdung compared to the normal compost of the same parental origin (Haritha Devi et al. 2009). *Pseudomonas, Paenibacillus, Azoarcus, Burkholderia, Spiroplasm, Acaligenes,* and *Acidobacterium,* the potential degraders of several categories of organics are seen associated with the earthworm's intestine and vermicasts (Singleton et al. 2003). Firmicutes viz., *Bacillus benzoevorans, B. cereus, B. licheniformis, B. megaterium, B. pumilus, B. subti-*

lis, B. macroides; Actinobacteria namely *Cellulosimicrobium cellulans, Microbacterium spp., M. oxydans*; Proteobacteria such as *Pseudomonas* spp., *P. libaniensis*; ungrouped genotypes *Sphingomonas* sp., *Kocuria palustris* and yeasts namely *Geotrichum* spp. and *Williopsis californica* were reported from vermicomposts (Vaz-Moreira et al. 2008). Pinel et al. (2008) reported the presence of a novel nephridial symbiont, *Verminephrobacter eiseniae* from *E. foetida. Ochrobactrum* sp., *Massilia* sp., *Leifsonia* sp. and bacteria belonging to families Aeromonadaceae, Comamonadaceae, Enterobacteriaceae, Flavobacteriaceae, Moraxellaceae, Pseudomonadaceae, Sphingobacteriaceae, Actinobacteria and Microbacteriaceae were reported to occur in earthworms alimentary canal (Byzov et al. 2009). The microbial flora of earthworm gut and cast are potentially active and can digest a wide range of organic materials and polysaccharides including cellulose, sugars, chitin, lignin, starch and polylactic acids Zhang et al. (2000; Aira et al. 2007; Vivas et al. 2009). Single-strand conformation polymorphism (SSCP) profiles on the diversity of eight bacterial groups viz., Alphaproteobacteria, Betaproteobacteria, Bacteroidetes, Gammaproteobacteria, Deltaproteobacteria, Verrucomicrobia, Planctomycetes, and Firmicutes from fresh soil, gut, and casts of the earthworms *L. terrestris* and *Aporrectodea caliginosa* showed the presence of Bacteroidetes, Alphaproteobacteria, Betaproteobacteria and representatives of classes Flavobacteria, Sphingobacteria (Bacteroidetes) and *Pseudomonas* spp. in the worm casts in addition to unclassified Sphingomonadaceae (Alphaproteobacteria) and *Alcaligenes* spp. (Betaproteobacteria) (Nechitaylo et al. 2010).

8.1.4 ROLE OF VERMICOMPOST IN SOIL FERTILITY

Vermicomposts can significantly influence the growth and productivity of plants (Kale et al. 1992; Kalembasa 1996; Edwards 1988; Sinha et al. 2009) due to their micro and macro elements, vitamins, enzymes and hormones (Makulec 2002). Vermicomposts contain nutrients such as nitrates, exchangeable phosphorus, soluble potassium, calcium, and magnesium in plant available forms (Orozco et al. 1996; Edwards 1998) and have large particular surface area that provides many microsites for microbial activity and for the strong retention of nutrients (Shi-wei and Fu-zhen 1991).

Uptake of nitrogen (N), phosphorus (P), potassium (K) and magnesium (Mg) by rice (*Oryza sativa*) plant was highest when fertilizer was applied in combination with vermicompost (Jadhav et al. 1997). N uptake by ridge gourd (*Luffa acutangula*) was higher when the fertilizer mix contained 50% vermicompost (Sreenivas et al. 2000). Apart from providing mineralogical nutrients, vermicomposts also contribute to the biological fertility by adding beneficial microbes to soil. Mucus, excreted through the earthworm's digestive canal, stimulates antagonism and competition between diverse microbial populations resulting in the production of some antibiotics and hormone-like biochemicals, boosting plant growth (Edwards and Bohlen 1996). In addition, mucus accelerates and enhances decomposition of organic matter composing stabilized humic substances which embody water-soluble phytohormonal elements (Edwards and Arancon 2004) and plant-available nutrients at high levels (Atiyeh et al. 2000c). Adding vermicasts to soil improves soil structure, fertility, plant growth and suppresses diseases caused by soil-borne plant pathogens, increasing crop yield (Chaoui et al. 2002; Scheuerell et al. 2005; Singh et al. 2008). Kale (1995) reported the nutrient status of vermicomposts with organic carbon 9.15-17.98%, total nitrogen 0.5-1.5%, available phosphorus 0.1-0.3%, available potassium 0.15%, calcium and magnesium 22.70-70 mg/100 g, copper 2–9.3 ppm, zinc 5.7-11.5 ppm and available sulphur 128–548 ppm.

Effects of a variety of vermicomposts on a wide array of field crops (Chan and Griffiths 1988; Arancon et al. 2004b), vegetable plants (Edwards and Burrows 1988; Wilson and Carlile 1989;Subler et al. 1998; Atiyeh et al. 2000b), ornamental and flowering plants (Edwards and Burrows 1988; Atiyeh et al. 2000c) under greenhouse and field conditions have been documented. Vermicomposts are used as alternative potting media due to their low-cost, excellent nutrient status and physiochemical characters. Considerable improvements in plant growth recorded after amending soils with vermicomposts have been attributed to the physico-chemical and biological properties of vermicomposts.

Vermicompost addition favorably affects soil pH, microbial population and soil enzyme activities (Maheswarappa et al. 1999) and also reduces the proportion of water-soluble chemical, which cause possible environmental contamination (Mitchell and Edwards 1997). Vermicompost addition increases the macropore space ranging from 50–500 μm, resulting in improved air-water relationship in the soil, favourably affecting plant growth (Marinari et al. 2000). Evaluation of various organic and inorganic amendments on growth of raspberry proves that vermicompost has beneficial buffering capability and ameliorate the damage caused by excess of nutrients which may otherwise cause phytotoxicity (Subler et al. 1998). Thus, vermicompost acts a soil conditioner (Albanell et al. 1988) and a slow-release fertilizer (Atiyeh et al. 2000a). During vermicomposting the heavy metals forms complex, aggregates with humic acids and other polymerized organic fractions resulting in lower availability of heavy metals to the plant, which are otherwise phytotoxic (Dominguez and Edwards 2004). Soil amended with vermicompost produced better quality fruits and vegetables with less content of heavy metals or nitrate, than soil fertilized with mineral fertilizers (Kolodziej and Kostecka 1994).

8.1.5 ROLE OF VERMICOMPOST BACTERIA IN BIOMEDICAL WASTE MANAGEMENT

The importance of sewage sludge, biosolids and biomedical waste management by safe, cheap and easy methods need no further emphasis. All these wastes are infectious and have to be disinfected before being disposed into the environment. Biosolids also contain an array of pathogenic microorganisms (Hassen et al. 2001). Biocomposting of wastes bring about biological transformation and stabilization of organic matter and effectively reduces potential risks of pathogens (Burge et al. 1987; Gliotti et al. 1997; Masciandaro et al. 2000). Vermicomposting does not involves a thermophilic phase which might increase the risk of using this technology for management of infectious wastes, but surprisingly vermicomposting resulted into a noticeable reduction in the pathogen indicators such as fecal coliform, Salmonella sp., enteric virus and helminth ova in the biosolids (Eastman 1999; Sidhu et al. 2001). Vermicomposting of biosolids resulted in reduction of faecal coliforms and *Salmonella* sp. from 39,000 MPN/g to 0 MPN/g and < 3 MPN to < 1MPN/g respectively (Dominguez and Edwards 2004). Vermicomposting of municipal sewage sludge with *L. mauritii* eliminated *Salmonella* and *Escherichia* sp., and the earthworm gut analysis also proved that *Salmonella* sp. ranging $15-17 \times 10^3$ CFU/g and *Escherichia* sp. ranging 10–14 × 10² CFU/g were completely eliminated in the gut after 70 days of vermicomposting period (Ganesh Kumar and Sekaran 2005). Activities by earthworms on sludge reduced levels of pathogens and odors of putrefaction and accelerated sludge stabilization (Mitchell 1978; Brown and Mitchell 1981; Hartenstein 1983). The reduction or removal of these enteric bacterial populations at the end of vermicomposting period, correlates with the findings that earthworm's diet include microorganisms and earthworms ability to selectively digest them (Bohlen and Edwards 1995; Edwards and Bohlen 1996). Apart from solid waste management, earthworms are also used in sewage water treatment. Earthworms promote the growth of 'beneficial decomposer bacteria' in wastewater and acts as aerators, grinders, crushers, chemical degraders, and biological stimulators (Dash 1978; Sinha et al. 2002). Earthworms also granulate the clay particles and increase the hydraulic conductivity and natural aeration and further grind the silt and sand particles and increase the total specific surface area and thereby enhance adsorption of the organic and inorganic matter from the wastewater. In addition, earthworms body acts as a 'biofilter' and remove the biological oxygen demand (BOD), chemical oxygen demand (COD), total dissolved solids (TDS) and total suspended solids (TSS) from wastewater by 90%, 80–90%, 90–92% and 90–95% respectively by 'ingestion' and biodegradation of organic wastes, heavy metals, and solids from wastewater and by their 'absorption' through body walls (Sinha et al. 2008).

Reports reveal that vermicomposting converts the infected biomedical waste containing various pathogens viz., *Staphylococcus aureus, Proteus vulgaris, Pseudomonas pyocyaneae* and *Escherichia coli* to an innocuous waste containing commensals like *Citrobactor freundii* and aerobic spore bearing microorganism usually found in the soil and alimentary canal of earthworms (Umesh et al. 2006). Vermicomposting plays a vital role for safe management of biomedical wastes and solid wastes generated from wastewater treatment plants and its bioconversion into valuable composts free from enteric bacterial populations. Depending on the earthworm species, vermicomposting was known to reduce the level of different pathogens such as *Salmonella enteriditis, Escherichia coli,* total and faecal coliforms, helminth ova and human viruses in different types of waste. Direct means of reduction in these microbial numbers during gut passage might be due to the digestive enzymes and mechanical grinding, while indirect means of pathogen removal might be due to promotion of aerobic conditions which could bring down the load of coliforms (Monroy et al. 2009; Edwards 2011; Aira et al. 2011).

8.1.6 ROLE OF VERMICOMPOST IN PLANT GROWTH PROMOTION

Use of vermicomposts as biofertilizers has been increasing recently due to its extraordinary nutrient status, and enhanced microbial and antagonistic activity. Vermicompost produced from different parent material such as food waste, cattle manure, pig manure, etc., when used as a media supplement, enhanced seedling growth and development, and increased productivity of a wide variety of crops (Edwards and Burrows 1988; Wilson and Carlile 1989; Buckerfield and Webster 1998; Edwards 1998; Subler et al. 1998; Atiyeh et al. 2000c). Vermicompost addition to soil-less bedding plant media enhanced germination, growth, flowering and fruiting of a wide range of green house vegetables and ornamentals (Atiyeh et al. 2000a, b, c), marigolds (Atiyeh et al. 2001), pepper (Arancon et al. 2003a), strawberries (Arancon et al. 2004b) and petunias (Chamani et al. 2008). Vermicompost application in the ratio of 20:1 resulted in a significant and consistent increase in plant growth in both field and greenhouse conditions (Edwards et al. 2004), thus providing a substantial evidence that biological growth promoting factors play a key role in seed germination and plant growth (Edwards and Burrows 1988; Edwards 1998). Investigations revealed that plant hormones and plant-growth regulating substances (PGRs) such as auxins, gibberellins, cytokinins, ethylene and abscisic acid are produced by microorganisms (Barea et al. 1976; Arshad and Frankenberger 1993).

Several researchers have documented the presence of plant growth regulators such as auxins, gibberellins, cytokinins of microbial origin (Krishnamoorthy and Vajranabhiah 1986; Grappelli et al. 1987; Tomati et al. 1988; Muscolo et al. 1999) and humic acids (Senesi et al. X1992; Masciandaro et al. 1997; Atiyeh et al. 2002) in vermicompost in appreciable quantities. Cytokinins produced by *Bacillus* and *Arthrobacter* spp.

in soils increase the vigour of seedlings (Inbal and Feldman 1982; Jagnow 1987). Microbially produced gibberellins influence plant growth and development (Mahmoud et al. 1984; Arshad and Frankenberger 1993) and auxins produced by *Azospirillum brasilense* affects the growth of plants belonging to paoceae (Barbieri et al. 1988). Extensive investigations on the biological activities of humic substances showed that they also posses plant growth stimulating property (Chen and Aviad 1990). Humic substances increased the dry matter yields of corn and oat seedlings (Lee and Bartlett 1976; Albuzio et al. 1994); number and length of tobacco roots (Mylonas and Mccants 1980); dry weights of roots, shoots and number of nodules of groundnut, soyabean and clover plants (Tan and Tantiwiramanond 1983) and vegetative growth of chicory plants (Valdrighi et al. 1996) and induced root and shoot formation in plant tissue culture (Goenadi and Sudharama 1995). High levels of humus have been reported from vermicomposts originating from food wastes, animal manure, sewage and paper mill sludges (Atiyeh et al. 2002; Canellas et al. 2002; Arancon et al. 2003c). The humic and fulvic acid in the humus dissolves insoluble minerals in the organic matter and makes them readily available to plants and in addition they also help plants to overcome stress and stimulates plant growth (Sinha et al. 2010). Studies on biological activities of vermicompost derived humic substances, revealed that they had similar growth-promoting hormonal effect (Dell'Agnola and Nardi 1987; Nardi et al. 1988; Muscolo et al. 1993). The humic materials extracted from vermicomposts have been reported to produce auxin-like cell growth and nitrate metabolism in carrots (*Daucus carota*) (Muscolo et al. 1996). Humates obtained from pig manure vermicompost increased growth of tomato (Atiyeh et al. 2002) and those obtained from cattle, food and paper waste vermicompost increased the growth of strawberries and peppers (Arancon et al. 2003a).

Earthworms produce plant growth regulators (Gavrilov 1963). Since earthworms increase the microbial activity by several folds they are considered as important agents which enhance the production of plant growth regulators (Nielson 1965; Graff and Makeschin 1980; Dell'Agnola and Nardi 1987; Grappelli et al. 1987; Tomati et al. 1987, 1988; Edwards and Burrows 1988; Nardi et al. 1988; Edwards 1998). Plant growth stimulating substances of microbial origin were isolated from tissues of *Aporrectodea longa, L. terrestris* and *Dendrobaena rubidus* and indole like substances were detected from the tissue extracts of *A. caliginosa, L. rubellus* and *E. foetida* which increased the growth of peas (Nielson 1965) and dry matter production of rye grass (Graff and Makeschin 1980). A. trapezoids aided in the dispersal of *Rhizobium* through soil resulting in increased root colonization and nodulation of leguminous plants (Bernard et al. 1994). Use of earthworm casts in plant propagation promoted root initiation, increased root numbers and biomass. The hormone-like effect produced by earthworm casts on plant metabolism, growth and development causing dwarfing, stimulation of rooting, internode elongation and precociousness of flowering was attributed to the fact of presence of microbial metabolites (Tomati et al. 1987; Edwards 1998). Earthworm casts stimulated growth of ornamental plants and carpophore formation in *Agaricus bisporus* when used as casing layer in mushroom cultivation (Tomati et al. 1987). Aqueous extracts of vermicompost produced growth comparable to the use of hormones such as auxins, gibberellins and cytokinins on Petunia, Begonia and Coleus, providing solid evidence that vermicompost is a rich source of plant growth regulating substances (Grappelli et al. 1987; Tomati et al. 1987, 1988). Addition of vermicompost at very low levels to the growth media dramatically increased the growth of hardy ornamentals *Chamaecyparis lawsonian, Elaeagnus pungens, Pyracantha* spp., *Viburnum bodnantense, Cotoneaster conspicus* and *Cupressocyparis leylandi.* Cucumber (Hahn and Bopp 1968), dwarf maize (Sembdner et al. 1976) and coleus bioassays (Edwards et al. 2004) evidenced that vermicompost contained appreciable amounts of cytokinins, gibberellins and auxins respectively. Maize seedlings dipped in vermicompost water showed marked difference in plumule length compared to normal water indicating that plant growth promoting hormones are present in vermicompost (Nagavallemma et al. 2004). Comparative studies on the impact of vermiwash and urea solution on seed germination, root and shoot length in *Cyamopsis tertagonoloba* proved that vermiwash contained hormone like substances (Suthar 2010). High performance liquid chromatography (HPLC) and gas chromatography-mass spectroscopy (GC-MS) analyses of aqueous extracts of cattle waste derived vermicompost showed presence of significant amounts indole-acetic-acid (IAA), gibberellins and cytokinins (Edwards et al. 2004).

Earthworm gut associated microbes enrich vermicomposts with highly water-soluble and light-sensitive plant growth hormones, which gets absorbed on humic acid substances in vermicompost making them extremely stable and helps them persist longer in soils thereby influencing plant growth (Atiyeh et al. 2002; Arancon et al. 2003c). This is confirmed by presence of exchangeable auxin group in the macrostructure of humic acid extract from vermicompost (Canellas et al. 2002). Apart from the rich nutritional status and ready nutrient availability, presence of humic acids and plant growth regulating substances makes vermicompost a biofertilizer which increases germination, growth, flowering and fruiting in a wide range of crops. Vermicompost substitution in a relatively small proportion (10–20%) to the potting mixture increased dry matter production and tomato growth significantly (Subler et al. 1998). Soil amended with 20% vermicompost was more suitable for tomato seedling production (Valenzuela et al. 1997). Similarly vermicompost addition upto 50% in the medium resulted in enhanced growth of *Chamaecyparis lawsoniana* (Lawson's Cypress), *Juniperus communis* (Juniper) and *Elaeagnus pungens* (Silverberry) rooted liners (Bachman and Edgar Davice 2000).

Vermicompost application increased plant spread (10.7%), leaf area (23.1%), dry matter (20.7%) and increased total strawberry fruit yield (32.7%) (Singh et al. 2008). Substitution of vermicompost drastically reduced the incidence of physiological disorders like albinism (16.1–4.5%), fruit malformation (11.5–4.0%) and occurrence of grey mould (10.4–2.1%) in strawberry indicating its significance in reducing nutrient-related disorders and Botrytis rot, thereby increasing the marketable fruit yield upto 58.6% with better quality parameters. Fruit harvested from plant receiving vermicompost were firmer, had higher total soluble solids (TSS), ascorbic acid content and attractive colour. All these parameters appeared to be dose dependent and best results were achieved at 7.5 t ha^{-1} (Singh et al. 2008). Vermicompost application showed significant increase in germination percent (93%), growth and yield of mung bean (*Vigna radiata*) compared to the control (Karmegam et al. 1999). Similarly, the fresh and dry matter yields of cowpea (*Vigna unguiculata*) were higher in soil amended with vermicompost than with biodigested slurry, (Karmegam and Daniel 2000). Combined application of vermicompost with N fertilizer gave higher dry matter (16.2 g plant^{-1}) and grain yield (3.6 t ha^{-1}) of wheat (*Triticum aestivum*) and higher dry matter yield (0.66 g plant^{-1}) of the following coriander (*Coriandrum sativum*) crop in wheat-coriander cropping system (Desai et al. 1999). Vermicompost application produced herbage yields of coriander cultivars comparable to those obtained with chemical fertilizers (Vadiraj et al. 1998). Yield of pea (*Pisum sativum*) increased with the application of vermicompost (10 t ha-1) along with recommended NPK (Meena et al. 2007). Vermicompost application to sorghum (*Sorghum bicolor*) (Patil and Sheelavantar 2000), sunflower (*Helianthus annuus*) (Devi et al. 1998), tomato (*Lycopersicon esculentum*) (Nagavallemma et al. 2004), eggplant (*Solanum melangona*) (Guerrero and Guerrero, 2006), okra (*Abelmoschus esculentus*) (Gupta et al. 2008), hyacinth bean (*Lablab purpureas*) (Karmegam and Daniel 2008), grapes (Buckerfield and Webster 1998) and cherry (Webster 2005) showed a positive result. Vermicompost amendment at the rate of 10 t ha[-1] along with 50% of recommended dose of NPK fertilizer increased the number and fresh weight of flowers per plant, flower diameter and yield, while at the rate of 15 t ha[-1] along with 50% of recommended dose of NPK increased vase life of *Chrysanthemum chinensis* (Nethra et al. 1999). Red Clover and cucumber grown in soil amended with vermicompost showed an increase in mineral contents viz., Ca, Mg, Cu, Mn and Zn in their shoot tissues (Sainz et al. 1998). Vermicomposted cow manure stimulated the growth of lettuce and tomato plants while the unprocessed parent material did not (Atiyeh et al. 2000b). Similarly, vermicomposted duck wastes resulted in better growth of tomatoes, lettuce, and peppers than the unprocessed wastes (Wilson and Carlile 1989). The enhancement in plant growth might be attributed to the fact that processed waste had improved physicochemical characteristics and nutrients, in forms readily available to the plant as well as the presence of plant growth promoting and antagonistic disease suppressing beneficial bacteria.

8.1.7 ROLE OF VERMICOMPOST IN PLANT DISEASE MANAGEMENT

8.1.7.1 PLANT PATHOGEN CONTROL

Soils with low organic matter and microbial activity are conducive to plant root diseases (Stone et al. 2004) and addition of organic amendments can effectively suppress plant disease (Raguchander et al. 1998; Blok et al.

2000; Lazarovits et al. 2000). Several researchers reported the disease suppressive properties of thermophilic compost (Hoitink et al. 1997; Goldstein 1998; Pitt et al. 1998) on a wide range of phytopathogens viz., *Rhizoctonia* (Kuter et al. 1983), *Phytopthora* (Hoitink and Kuter 1986; Pitt et al. 1998), *Plasmidiophora brassicae* and *Gaeumannomyces graminis* (Pitt et al. 1998) and *Fusarium* (Kannangowa et al. 2000; Cotxarrera et al. 2002). Microbial antagonism might be one of the possible reasons for disease suppression as organic amendments enhances the microbial population and diversity. Traditional thermophilic composts promote only selected microbes while non-thermophilic vermicomposts are rich sources of microbial diversity and activity and harbour a wide variety of antagonistic bacteria thus acts as effective biocontrol agents aiding in suppression of diseases caused by soil-borne phytopathogenic fungi (Chaoui et al. 2002; Scheuerell et al. 2005; Singh et al. 2008).

Earthworm feeding reduces the survival of plant pathogens such as *Fusarium* sp. and *Verticillium dahliae* (Yeates 1981; Moody et al. 1996) and increases the densities of antagonistic fluorescent pseudomonads and filamentous actinomycetes while population densities of *Bacilli* and *Trichoderma* spp. remains unaltered (Elmer 2009). Earthworm activities reduce root diseases of cereals caused by *Rhizoctonia* (Doube et al. 1994). It has been proved that earthworms decreased the incidence of field diseases of clover, grains, and grapes incited by *Rhizoctonia* spp. (Stephens et al. 1994a; Stephens and Davoren 1997) and *Gaeumannomyces* spp. (Clapperton et al. 2001). Earthworms *Aporrectodea trapezoides* and *Aporrectodea rosea* act as vectors of *Pseudomonas* corrugata 214OR, a biocontrol agent for wheat take-all caused by *G. graminis* var. *tritd* (Doube et al. 1994). Greenhouse studies on augmentation of pathogen infested soils with *L. terrestris* showed a significant reduction of disease caused by *Fusarium oxysporum* f. sp. *asparagi* and *F. proliferatum* on susceptible cultivars of asparagus (*Asparagus officinalis*), *Verticillium dahliae* on eggplant (*Solanum melongena*) and *F. oxysporum* f. sp. *lycopersici* race 1 on tomato. Plant weights increased by 60-80% and disease severity reduced by 50-70% when soils were augmented with earthworms. Incorporation of soil with vermicompost effectively suppressed *R. solani* in wheat (Stephens et al. 1993), *Phytophthora nicotianae* (Nakamura 1996; Szczech 1999; Szczech and Smolinska 2001) and *Fusarium* in tomatoes (Nakamura

1996; Szczech 1999), *Plasmodiophora brassicae* in tomatoes and cabbage (Nakamura 1996), *Pythium* and *Rhizoctonia* (root rot) in cucumber and radish (Simsek Ersahin et al. 2009), *Botrytis cineria* (Singh et al. 2008) and *Verticillium* (Chaoui et al. 2002) in strawberry and S*phaerotheca fulginae* in grapes (Edwards et al. 2004). Vermicompost application drastically reduced the incidence of 'Powdery Mildew', 'Color Rot' and 'Yellow Vein Mosaic' in Lady's finger (*Abelmoschus esculentus*) (Agarwal et al. 2010). Substitution of vermicompost in the growth media reduced the fungal diseases caused by *R. solani, P. drechsleri* and *F. oxysporum* in gerbera (Rodriguez et al. 2000). Amendment of vermicompost at low rates (10-30%) in horticulture bedding media resulted in significant suppression of *Pythium* and *Rhizoctonia* under green house conditions (Edwards et al. 2004). Research findings proved that vermicompost when added to container media significantly reduced the infection of tomato plants by *P. nicotianae* var. *nicotianae* and *F. oxysporum* sp. *lycopersici* (Szech et al. 1993; Szczech 1999). Club-rot of cabbage caused by *P. brassicae* was inhibited by dipping cabbage roots into a mixture of clay and vermicompost (Szech et al. 1993). Potato plants treated with vermicompost were less susceptible to *P. infestans* than plants treated with inorganic fertilizers (Kostecka et al. 1996a). Aqueous extracts of vermicompost inhibited mycelial growth of *B. cineria, Sclerotinia sclerotiorum, Corticium rolfsii, R. solani* and *F. oxysporum* (Nakasone et al. 1999), effectively controlled powdery mildew of barley (Weltzien 1989) and affected the development of powdery mildews on balsam (*Impatiens balsamina*) and pea (*Pisum sativum*) caused by *Erysiphe cichoracearum* and *E. pisi*, respectively in field conditions (Singh et al. 2003).

8.1.7.2 MECHANISMS THAT MEDIATE PATHOGEN SUPPRESSION

Two possible mechanisms of pathogen suppression have been described, one depends on systemic plant resistance and the other is mediated by microbial competition, antibiosis and hyperparasitism (Hoitink and Grebus 1997). The microbially mediated suppression is again classified into two mechanisms viz., 'general suppression' where a wide range of microbes suppress the pathogens such as *Pythium* and *Phytopthora* (Chen et al. 1987) and 'specific suppression' where a narrow range of organisms facilitates suppression, for instance disease caused by *Rhizoctonia* (Hoitink et al. 1997). The disease suppressive effect of vermicompost against fusarium wilt of tomato clearly depicted that fungus inhibition was purely biotic and no chemical factors played any role, since the experiments with heat-sterilized vermicompost failed to control the disease (Szczech 1999). Experiments on suppression of damping-off caused by *R. solani*, in vermicompost amended nurseries of white pumpkin proved that vermicompost suppressed the disease in a dosage and temperature dependent manner (Rivera et al. 2004). Earthworm castings are rich in nutrients (Lunt and Jacobson 1944; Parle 1963) and calcium humate, a binding agent (Edwards 1998) that reduces desiccation of individual castings and favors the incubation and proliferation of beneficial microbes, such as *Trichoderma* spp. (Tiunov and Scheu 2000), *Pseudomonas* spp. (Schmidt et al. 1997), and mycorrhizal spores (Gange 1993; Doube et al. 1995). Earthworm activity increased the communities of Gram-negative bacteria (Clapperton et al. 2001; Elmer 2009). Vermicompost associated chitinolytic bacterial communities viz., *Nocardioides oleivorans*, several species of *Streptomyces* and *Staphylococcus epidermidis* showed inhibitory effects against plant phytopathogens such as, *R. solani*, *Colletotrichum coccodes*, *Pythium ultimum*, *P. capsici* and *Fusarium moniliforme* (Yasir et al. 2009a).

8.1.8 ROLE OF VERMICOMPOST IN ARTHROPOD PEST CONTROL

Addition of organic amendments helped in suppression of various insect pests such as European corn borer (Phelan et al. 1996), other corn insect pests (Biradar et al. 1998), aphids and scale insects (Culliney and Pimentel 1986; Costello and Altiei 1995; Huelsman et al. 2000) and brinjal shoot and fruit borer (Sudhakar et al. 1998). Several reports also evidenced that vermicompost addition decreased the incidence of *Spodoptera litura*, *Helicoverpa armigera*, leaf miner (*Apoaerema modicella*), jassids (*Empoasca kerri*), aphids (*Aphis craccivora*) and spider mites on groundnuts (Rao et al. 2001; Rao 2002, 2003) and psyllids (*Heteropsylla cubana*) on a tropical leguminous tree (*Leucaena leucocephala*) (Biradar et al. 1998).

Vermicompost amendment decreased the incidence of sucking pests under field conditions (Ramesh 2000) and suppressed the damage caused by of two-spotted spider mite (*Tetranychus* spp.), aphid (*Myzus persicae*) (Edwards et al. 2007) and mealy bug (*Pseudococcus* spp.) under green house conditions (Arancon et al. 2007). Vermicompost substitution to soil less plant growth medium MetroMix 360 (MM360) at a rate less then 50% reduced the damage caused by infestation of pepper seedlings by *M. persicae* and *Pseudococcus* spp. and tomato seedlings by *Pseudococcus* spp., cabbage seedlings by *M. persicae* and cabbage white caterpillars (*Pieris brassicae* L.) (Arancon et al. 2005). Greenhouse cage experiments conducted on tomatoes and cucumber seedlings infested with *M. persicae*, citrus mealybug (*Planococcus citri*), two spotted spider mite (*Tetranychus urticae*); striped cucumber beetles (*Acalymna vittatum*) attacking cucumbers and tobacco hornworms (*Manduca sexta*) attacking tomatoes proved that treatment of infested plants with aqueous extracts of vermicompost suppressed pest establishment, and their rates of reproduction. Vermicompost teas at higher dose also brought about pest mortality (Edwards et al. 2010b). Suppression of aphid population gains importance since they are key vectors in transmission of plant viruses. Addition of solid vermicompost reduced damage by *A. vittatum* and spotted cucumber beetles (*Diabotrica undecimpunctata*) on cucumbers and larval hornworms (*Manduca quinquemaculata*) on tomatoes in both greenhouse and field experiments (Yardim et al. 2006). Combined application of vermicompost and vermiwash spray to chilli (*Capiscum annum*) significantly reduced the incidence of 'Thrips' (*Scirtothrips dorsalis*) and 'Mites' (*Polyphagotarsonemus latus*) (Saumaya et al. 2007).

8.1.8.1 MECHANISMS THAT MEDIATE PEST CONTROL

Plants grown in inorganic fertilizers are more prone to pest attack than those grown on organic fertilizers (Culliney and Pimentel 1986; Yardim and Edwards 2003; Phelan 2004). Inorganic nitrogen fertilization improves the nutritional quality and palatability of the host plants, inhibits the raise of secondary metabolite concentrations (Fragoyiannis et al. 2001; Herms 2002), enhances the fecundity of insects dieting on them, attracts more individuals for oviposition (Bentz et al. 1995) and increases the population growth rates of insects (Culliney and Pimentel 1986; Jannsson and Smilowitz 1986). Though organic fertilizer has an enhanced nutritional composition they release nutrients at a slower rate (Patriquin et al. 1995) hence plants grown with organic fertilizers possess decreased N levels (Steffen et al. 1995) and have higher phenol content (Asami et al. 2003) resulting in resistance of these plants to pest attack. Similarly vermicomposts exhibit a slow, balanced nutritional release pattern, particularly in release of plant available N, soluble K, exchangeable Ca, Mg and P (Edwards and Fletcher 1988; Edwards 1998). Vermicomposts are rich in humic acid and phenolic compounds. Phenolic compounds act as feeding deterrents and hence significantly affect pest attacks (Kurowska et al. 1990; Summers and Felton 1994; QiTian 2004; Hawida et al. 2007; Koul 2008; Mahanil et al. 2008; Bhonwong et al. 2009). Soil containing earthworms contained polychlorinated phenols and their metabolites (Knuutinen et al. 1990). An endogenous phenoloxidase present in *L. rubellus* bioactivate compounds to form toxic phenols viz., p-nitrophenol (Park et al. 1996). Monomeric phenols could be absorbed by humic acids in the gut of earthworms (Vinken et al. 2005). Uptake of soluble phenolic compounds from vermicompost, by the plant tissues makes them unpalatable thereby affecting pest rates of reproduction and survival (Edwards et al. 2010a; Edwards et al. 2010b).

8.1.9 ROLE OF VERMICOMPOST IN NEMATODE CONTROL

It has been well documented that addition of organic amendments decreases the populations of plant parasitic nematodes (Addabdo 1995; Sipes et al. 1999; Akhtar and Malik 2000). Vermicompost amendments appreciably suppress plant parasitic nematodes under field conditions (Arancon et al. 2003b). Vermicomposts also suppressed the attack of Meloidogyne incognita on tobacco, pepper, strawberry and tomato (Swathi et al. 1998; Edwards et al. 2007; Arancon et al. 2002; Morra et al. 1998) and decreased the numbers of galls and egg masses of *Meloidogyne javanica* (Ribeiro et al. 1998).

8.1.9.1 MECHANISMS THAT MEDIATE NEMATODE CONTROL

There are several feasible mechanisms that attribute to the suppression of plant parasitic nematodes by vermicompost application and it involves both biotic and abiotic factors. Organic matter addition to the soil stimulates the population of bacterial and fungal antagonists of nematodes (e.g., *Pasteuria penetrans*, *Pseudomonas* spp. and chitinolytic bacteria, *Trichoderma* spp.), and other typical nematode predators including nematophagous mites viz., *Hypoaspis calcuttaensis* (Bilgrami 1996), Collembola and other arthropods which selectively feeds on plant parasitic nematodes. (Thoden et al. 2011). Vermicompost amendment promoted fungi capable of trapping nematode and destroying nematode cysts (Kerry 1988) and increased the population of plant growth-promoting rhizobacteria which produce enzymes toxic to plant parasitic nematodes (Siddiqui and Mahmood 1999). Vermicompost addition to soils planted with tomatoes, peppers, strawberry and grapes showed a significant reduction of plant parasitic nematodes and increased the population of fungivorous and bacterivorous nematodes compared to inorganic fertilizer treated plots (Arancon et al. 2002). In addition, few abiotic factors viz., nematicidal compounds such as hydrogen sulphide, ammonia, nitrates, and organic acids released during vermicomposting, as well as low C/N ratios of the compost cause direct adverse effects while changes in soil physiochemical characterists viz., bulk density, porosity, water holding capacity, pH, EC, CEC and nutrition posses indirect adverse effects on plant parasitic nematodes (Rodriguez-Kabana 1986; Thoden et al. 2011).

8.2 CONCLUSION

Vermicomposting is a cost-effective and eco-friendly waste management technology which takes the privilege of both earthworms and the associated microbes and has many advantages over traditional thermophilic composting. Vermicomposts are excellent sources of biofertilizers and their addition improves the physiochemical and biological properties of agricultural soil. Vermicomposting amplifies the diversity and population of beneficial microbial communities. Although there are some reports indicating that few harmful microbes such as spores of *Pythium* and *Fusarium* are dispersed by earthworms (Edwards and Fletcher 1988), the presence and amplification of antagonistic disease-suppressing and other plant growth-promoting beneficial bacteria during vermicomposting out weigh these harmful effects (Edwards and Fletcher 1988; Gammack et al. 1992; Brown 1995). Vermicomposts with excellent physio-chemical properties and buffering ability, fortified with all nutrients in plant available forms, antagonistic and plant growth-promoting bacteria are fantabulous organic amendments that act as a panacea for soil reclamation, enhancement of soil fertility, plant growth, and control of pathogens, pests and nematodes for sustainable agriculture.

REFERENCES

1. Addabdo TD (1995) The nematicidal effect of organic amendments: a review of the literature 1982–1994. Nematol Mediterranea 23:299-305
2. Aira M, Monroy F, Dominguez J (2007) Earthworms strongly modify microbial biomass and activity triggering enzymatic activities during vermicomposting independently of the application rates of pig slurry. Sci Total Environ 385:252-261
3. Aira M, Gómez-Brandón M, González-Porto P, Domínguez J (2011) Selective reduction of the pathogenic load of cow manure in an industrial-scale continuous-feeding vermireactor. Bioresource Technol 102:9633-9637
4. Agarwal S, Sinha RK, Sharma J, et al. (2010) Ver-miculture for sustainable horticulture: Agronomic impact studies of earthworms, cow dung compost and vermi-compost vis-à-vis chemical fertilizers on growth and yield of lady's finger (Abelmoschus esculentus). In: Sinha RK (ed) Special Issue on 'Vermiculture Technology', International Journal of Environmental Engineering, Inderscience Publishers, Geneva, Switzerland.
5. Akhtar M, Malik A (2000) Role of organic amendments and soil organisms in the biological control of plant parasitic nematodes: a review. Bioresour Technol 74:35-47
6. Albanell E, Plaixats J, Cabrero T (1988) Chemical changes during vermicomposting (Eisenia fetida) of sheep manure mixed with cotton industrial wastes. Biol Fertil Soils 6:266-269
7. Albuzio A, Concheri G, Nardi S, Dell'Agnola G (1994) Effect of humic fractions of different molecular size on the development of oat seedlings grown in varied nutritional conditions. In: Senesi N, Miano TM (eds) Humic substances in the Global Environment and Implications on Human Health, Elsevier, Amsterdam, Netherlands. pp 199-204

8. Arancon NQ, Edwards CA, Atiyeh R, Metzger JD (2004) Effects of vermicomposts produced from food waste on the growth and yields of greenhouse peppers. Bioresour Technol 93:139-144
9. Arancon NQ, Edwards CA, Bierman P, Metzger JD, Lee S, Welch C (2003) Effects of vermicomposts to tomatoes and peppers grown in the field and strawberries under high plastic tunnels. Pedobiologia 47:731-735
10. Arancon NQ, Edwards CA, Bierman P, Welch C, Metzger JD (2004) The influence of vermicompost applications to strawberries: Part 1. Effects on growth and yield. Bioresour Technol 93:145-153
11. Arancon NQ, Edwards CA, Lee S (2002) Management of plant parasitic nematode populations by use of vermicomposts. Proceedings Brighton Crop Protection Conference – Pests and Diseases. 705-716
12. Arancon NQ, Edwards CA, Yardim EN, Oliver TJ, Byrne RJ, Keeney G (2007) Suppression of two-spotted spider mite (Tetranychus urticae), mealy bug (Pseudococcus sp) and aphid (Myzus persicae) populations and damage by vermicomposts. Crop Prot 26:29-39
13. Arancon NQ, Galvis PA, Edwards CA (2005) Suppression of insect pest populations and damage to plants by vermicomposts. Bioresour Technol 96:1137-1142
14. Arancon NQ, Galvis P, Edwards CA, Yardim E (2003) The trophic diversity of nematode communities in soils treated with vermicomposts. Pedobiologia 47:736-740
15. Arancon NQ, Lee S, Edwards CA, Atiyeh RM (2003) Effects of humic acids and aqueous extracts derived from cattle, food and paper-waste vermicomposts on growth of greenhouse plants. Pedobiologia 47:744-781
16. Arshad M, Frankenberger WT Jr (1993) Microbial production of plant growth regulators. In: Metting FB Jr (ed) Soil Microbial Ecology: Applications in Agricultural and Environmental Management, Marcell Dekker, New York. pp 307-347
17. Asami DK, Hang YJ, Barnett DM, Mitchell AE (2003) Comparison of the total phenolic and ascorbic acid content of freeze-dried and air-dried marionberry, strawberry and corn grown using conventional organic and sustainable agricultural practices. J Agric Food Chem 51:1237-1241
18. Asha A, Tripathi AK, Soni P (2008) Vermicomposting: A Better Option for Organic Solid Waste Management. J Hum Ecol 24:59-64
19. Atiyeh RM, Arancon NQ, Edwards CA, Metzger JD (2001) The influence of earthworm-processed pig manure on the growth and productivity of marigolds. Bioresour Technol 81:103-108
20. Atiyeh RM, Dominguez J, Subler S, Edwards CA (2000) Changes in biochemical properties of cow manure during processing by earthworms (Eisenia andrei, Bouché) and the effects on seedling growth. Pedobiologia 44:709-724
21. Atiyeh RM, Lee S, Edwards CA, Arancon NQ, Metzger JD (2002) The influence of humic acids derived from earthworm-processed organic wastes on plant growth. Bioresour Technol 84:7-14
22. Atiyeh RM, Subler S, Edwards CA, Bachman G, Metzger JD, Shuster W (2000) Effects of vermicomposts and composts on plant growth in horticulture container media and soil. Pedobiologia 44:579-590

23. Atiyeh RM, Arancon NQ, Edwards CA, Metzger JD (2000) Influence of earthworm-processed pig manure on the growth and yield of green house tomatoes. Bioresour Technol 75:175-180

24. Ayyadurai N, Ravindra Naik P, Sakthivel N (2007) Functional characterization of antagonistic fluorescent pseudomonads associated with rhizospheric soil of rice (Oryza sativa L.). J Microbiol Biotechnol 17:919-927

25. Bachman GR, Edgar Davice W (2000) Growth of magnolia virginiana liners in vermicompost-amended media. Proceeding of SNA Research Conference, Southern Nursery Association, Atlanta. pp 65-67

26. Bansal S, Kapoor KK (2000) Vermicomposting of crop residues and cattle dung with Eisenia foetida. Bioresour Technol 73:95-98

27. Banu JR, Logakanthi S, Vijayalakshmi GS (2001) Biomanagement of paper mill sludge using an indigenous (Lampito mauritii) and two exotic (Eudrilus eugineae and Eisenia foetida) earthworms. J Environ Biol 22:181-185

28. Barbieri P, Bernardi A, Galli E, Zanetti G (1988) Effects of inoculation with different strains of Azospirillum brasilense on wheat roots development. In: Klingmüller W (ed) Azospirillum IV, Genetics, Physiology, Ecology, SpringerVerlag, Berlin. pp 181-188

29. Barea JM, Navarro E, Montana E (1976) Production of plant growth regulators by rhizosphere phosphate solubilizing bacteria. J Appl Bacteriol 40:129-134

30. Benitez E, Nogales R, Elvira C, Masciandaro G, Ceccanti B (1999) Enzymes activities as indicators of the stabilization of sewage sludges composting by Eisenia foetida. Bioresour Technol 67:297-303

31. Bentz JA, Reeves J, Barbosa P, Francis B (1995) Nitrogen fertilizer effect on selection, acceptance and suitability of Euphorbia pulcherrima (Euphorbiaceae) as a host plant to Bemisia tabaci (Homoptera: Aleyrodidae). Environ Entomol 24:40-45

32. Bernal MP, Faredes C, Sanchez-Monedero MA, Cegarra J (1998) Maturity and stability parameters of composts prepared with a, wide range of organic wastes. Bioresour Technol 63:91-99

33. Bernard MD, Peter MS, Christopher WD, Maarten HR (1994) Interactions between earthworms, beneficial soil microorganisms and root pathogens. Appl Soil Ecol 1:3-10

34. Bhatnagar RK, Palta RK (1996) Earthworm-Vermiculture and Vermicomposting. Kalyani Publishers, New Delhi.

35. Bhonwong A, Stout MJ, Attajarusit J, Tantasawat P (2009) Defensive role of tomato polyphenoloxidases against cotton bollworm (Helicoverpa armigera) and beet army worm (Spodoptera exigua). J Chem Ecol 35:28-38

36. Bilgrami AL (1996) Evaluation of the predation abilities of the mite Hypoaspis calcuttaensis, predaceous on plant and soil nematodes. Fund Appl Nematol 20:96-98

37. Binet F, Fayolle L, Pussard M (1998) Significance of earthworms in stimulating soil microbial activity. Biol Fertil Soils 27:79-84

38. Biradar AP, Sunita ND, Teggel RG, Devaradavadgi SB (1998) Effect of vermicompost on the incidence of subabul psyllid. Insect- Environ 4:55-56

39. Blok WJ, Lamers JG, Termoshuizen AJ, Bollen GJ (2000) Control of soil-borne plant pathogens by incorporating fresh organic amendments followed by tarping. Phytopathology 90:253-259

40. Bohlen PJ, Edwards CA (1995) Earthworm effects on N dynamics and soil respiration in microcosms receiving organic and inorganic nutrients. Soil Biol Biochem 27:341-348

41. Brown BA, Mitchell MJ (1981) Role of the earthworm, Eisenia fetida, in affecting survival of Salmonella enteritidis ser. Typhimurium. Pedobiologia 22:434-438

42. Brown GG (1995) How do earthworms affect microfloral and faunal community diversity? Plant Soil 170:209-231

43. Brown GG, Barois I, Lavelle P (2000) Regulation of soil organic matter dynamics and microbial activity in the drilosphere and the role of interactions with other edaphic functional domains. Eur J Soil Biol 36:177-198

44. Buckalew DW, Riley RK, Yoder WA, Vail WJ (1982) Invertebrates as vectors of endomycorrhizal fungi and Rhizobium upon surface mine soils. West Virginia Acad Sci Proc 54:1

45. Buckerfield JC, Webster KA (1998) Worm-worked waste boosts grape yields: prospects for vermicompost use in vineyards. Australian and New Zealand Wine Industry Journal 13:73-76

46. Burge WD, Enkiri NK, Hussong D (1987) Salmonella regrowth in compost as influenced by substrate. Microbial Ecol 14:243-253

47. Butt KR (1993) Utilization of solid paper mill sludge and spent brewery yeast as a feed for soil dwelling earthworms. Bioresour Technol 44:105-107

48. Byzov BA, Nechitaylo TY, Bumazhkin BK, Kurakov AV, Golyshin PN, Zvyagintsev DG (2009) Culturable microorganisms from the earthworm digestive tract. Microbiology 78:360-368

49. Canellas LP, Olivares FL, Okorokova FAR (2002) Humic acids isolated from earthworm compost enhance root elongation, lateral root emergence and plasma membrane H+− ATPase activity in maize roots. Plant Physiol 130:1951-1957

50. Chamani E, Joyce DC, Reihanytabar A (2008) Vermicompost Effects on the Growth and Flowering of Petunia hybrida 'Dream Neon Rose'. American-Eurasian J Agric And Environ Sci 3:506-512

51. Chan LPS, Griffiths DA (1988) The vermicomposting of pretreated pig manure. Biol Wastes 24:57-69

52. Chaoui H, Edwards CA, Brickner M, Lee S, Arancon N (2002) Suppression of the plant diseases, Pythium (damping off), Rhizoctonia (root rot) and Verticillum (wilt) by vermicomposts. Proceedings of Brighton Crop Protection Conference – Pests and Diseases II(8B-3):711-716

53. Chen W, Hoitink HA, Schmitthenner AF, Touvinen O (1987) The role of microbial activity in suppression of damping off caused by Pythium ultimum. Phytopathology 78:314-322

54. Chen Y, Aviad T (1990) Effects of humic substances on plant growth. In: MacCarthy P, Clapp CE, Malcolm RL, Bloom PR (eds) Humic Substances in Soil and Crop Sciences, Selected Reading ASA and SSSA, Madison. pp 161-186

55. Citernesi U, Neglia R, Seritti A, Lepidi AA, Filippi C, Bagnoli G, Nuti MP, Galluzzi R (1977) Nitrogen fixation in the gastro-enteric cavity of soil animals. Soil Biol Biochem 9:71-72

56. Clapperton MJ, Lee NO, Binet F, Conner RL (2001) Earthworms indirectly reduce the effect of take-all (Gaeumannomyces graminis var. tritici) on soft white spring wheat (Triticium aestivum cv. Fielder). Soil Biol Biochem 33:1531-1538

57. Contreras E (1980) Studies on the intestinal actinomycete flora of Eisenia lucens (Annelida, Oligochaeta). Pedobiologia 20:411-416

58. Correa JD, Barrios ML, Galdona RP (2004) Screening for plant growth promoting rhizobacteria in Chamaecytisus proliferus (tagasaste), a forage tree-shrub legume endemic to the Canary Islands. Plant Soil 266:75-84

59. Costello MJ, Altiei MA (1995) Abundance, growth rate and parasitism of Brevicoryne brassicae and Myzus persicae (Homoptera: Aphididae) on broccoli grown in living mulches. Agric Ecosyst Environ 52:187-196

60. Cotxarrera L, Trillas-Gayl MI, Steinberg C, Alabouvette C (2002) Use of sewage sludge compost and Trichoderma asperellum isolates to suppress Fusarium wilt of tomato. Soil Biol Biochem 34:467-476

61. Culliney TW, Pimentel D (1986) Ecological effects of organic agricultural practices on insect populations. Agric Ecosyst Environ 15:253-256

62. Daniel O, Anderson JM (1992) Microbial biomass and activity in contrasting soil material after passage through the gut of earthworm Lumbricus rubellus Hoffmeister. Soil Biol Biochem 24:465-470

63. Darwin F, Seward AC (1903) More letters of Charles Darwin. In: John M (ed) A record of his work in series of hitherto unpublished letters, London. p 508

64. Dash MC (1978) Role of earthworms in the decomposer system. In: Singh JS, Gopal B (eds) Glimpses of ecology, India International Scientific Publication, New Delhi. pp 399-406

65. Dell'Agnola G, Nardi S (1987) Hormone-like effect and enhanced nitrate uptake induced by depolyconder humic fractions obtained from Allolobophora rosea and A. caliginosa faeces. Biol Fertil Soils 4:111-118

66. Desai VR, Sabale RN, Raundal PV (1999) Integrated nitrogen management in wheat-coriander cropping system. J Maharasthra Agric Univ 24:273-275

67. Devi D, Agarwal SK, Dayal D (1998) Response of sunflower (Helianthus annuus) to organic manures and fertilizers. Indian J Agron 43:469-473

68. Devliegher W, Verstraete W (1995) Lumbricus terrestris in a soil core experiment: nutrient-enrichment processes (NEP) and gut-associated processes (GAP) and their effect on microbial biomass and microbial activity. Soil Biol Biochem 27:1573-1580

69. Devliegher W, Verstraete W (1997) Microorganisms and soil physicochemical conditions in the drilosphere of Lumbricus terrestris. Soil Biol Biochem 29:1721-1729

70. Dominguez J, Edwards CA, Webster M (2000) Vermicomposting of sewage sludge: effects of bulking materials on the growth and reproduction of the earthworm Eisenia andrei. Pedobiologia 44:24-32

71. Dominguez J, Edwards CA (2004) Vermicomposting organic wastes: A review. In: Shakir Hanna SH, Mikhail WZA (eds) Soil Zoology for sustainable Development in the 21st century, Cairo. pp 369-395

72. Doube BM, Ryder MH, Davoren CW, Meyer T (1995) Earthworms: a down under delivery system service for biocontrol agents of root disease. Acta Zool Fennica 196:219-223

73. Doube BM, Stephens PM, Davorena CW, Ryderb MH (1994) Interactions between earthworms, beneficial soil microorganisms and root pathogens. Appl Soil Ecol 1:3-10

74. Eastman BR (1999) Achieving pathogen stabilization using vermicomposting. Bio-Cycle 40:62-64

75. Edwards CA (1988) Breakdown of animal, vegetable and industrial organic wastes by earthworms. In: Edwards CA, Neuhauser EF (eds) Earthworms in Waste and Environmental Management SPB, The Hague, Netherlands. pp 21-31

76. Edwards CA (1998) The use of earthworms in the breakdown and management of organic wastes. In: Edwards CA (ed) Earthworm Ecology, CRC Press, Boca Raton. pp 327-354

77. Edwards CA (2011) Human pathogen reduction during vermicomposting. In: Edwards CA, Arancon NQ, Sherman R (eds) Vermiculture technology: earthworms, organic wastes and environmental management, CRC Press, Boca Raton. pp 249-261

78. Edwards CA, Arancon NQ, Bennett MV, Askar A, Keeney G, Little B (2010) Suppression of green peach aphid (Myzus persicae) (Sulz.), citrus mealybug (Planococcus citri) (Risso), and two spotted spider mite (Tetranychus urticae) (Koch.) attacks on tomatoes and cucumbers by aqueous extracts from vermicomposts. Crop Prot 29:80-93

79. Edwards CA, Arancon NQ, Bennett MV, Askar A, Keeney G (2010) Effect of aqueous extracts from vermicomposts on attacks by cucumber beetles (Acalymna vittatum) (Fabr.) on cucumbers and tobacco hornworm (Manduca sexta) (L.) on tomatoes. Pedobiologia 53:141-148

80. Edwards CA, Arancon NQ, Emerson E, Pulliam R (2007) Supressing plant parasitic nematodes and arthropod pests with vermicompost teas. Biocycle. 38-39

81. Edwards CA, Arancon NQ (2004) Vermicomposts suppress plant pest and disease attacks. BioCycle 45:51-53

82. Edwards CA, Bohlen PJ (1996) Biology and Ecology of earthworms. Chapman and Hall, London. p 426

83. Edwards CA, Burrows I, Fletcher KE, Jones BA (1985) The use of earthworms for composting farm wastes. In: Gasser JKR (ed) Composting Agricultural and Other Wastes, Elsevier, London. pp 229-241

84. Edwards CA, Burrows I (1988) The potential of earthworm composts as plant growth media. In: Edwards CA, Neuhauser E (eds) Earthworms in Waste and Environmental Management, SPB Academic Press, The Hague. pp 21-32

85. Edwards CA, Dominguez J, Arancon NQ (2004) The influence of vermicomposts on pest and diseases. In: Shakir Hanna SH, Mikhail WZA (eds) Soil Zoology for Sustainable Development in the 21st centuary, Cairo. pp 397-418

86. Edwards CA, Dominguez J, Neuhauser EF (1998) Growth and reproduction of Perionyx excavatus (Perr.) (Megascolecidae) as factors in organic waste management. Biol Fertil Soils 27:155-161

87. Edwards CA, Fletcher KE (1988) Interaction between earthworms and microorganisms in organic matter breakdown. Agric Ecosyst Environ 20:235-249

88. Edwards CA, Lofty R (1977) The Biology of Earthworms. Chapmann and Hall, London.

89. Elmer WH (2009) Influence of earthworm activity on soil microbes and soilborne diseases of vegetables. Plant Dis 93:175-179
90. Elvira C, Dominguez J, Sampedro L, Mato S (1995) Vermicomposting for the pulp industry. Biocycle 36:62-63
91. Elvira C, Sampedro L, Benítez E, Nogales R (1998) Vermicomposting of sludges from paper mill and dairy industries with Eisenia andrei: a pilot-scale study. Bioresour Technol 63:205-211
92. Elvira C, Sampedro L, Dominguez J, Mato S (1997) Vermicomposting of wastewater sludge from paper-pulp industry with nitrogen rich materials. Soil Biol Biochem 9:759-762
93. Fracchia L, Dohrmann AB, Martinotti MG, Tebbe CC (2006) Bacterial diversity in a finished compost and vermicompost: differences revealed by cultivation independent analyses of PCR-amplified 16S rRNA genes. Appl Microbiol Biotechnol 71:942-952
94. Fragoyiannis DA, McKinlay RG, D'Mello JPF (2001) Interactions of aphids herbivory and nitrogen availability on the total foliar glycoalkoloid content of potato plants. J Chem Ecol 27:1749-1762
95. Gajalakshmi S, Ramasamy EV, Abbasi SA (2001) Assessment of sustainable vermiconversion of water hyacinth at different reactor efficiencies employing Eudrilus engeniae Kingburg. Bioresour Technol 80:131-135
96. Gajalakshmi S, Ramasamy EV, Abbasi SA (2002) Vermicomposting of paper waste with the anecic earthworm Lampito mauritii Kingburg. Indian J Chem Technol 9:306-311
97. Gammack SM, Paterson E, Kemp JS, Cresser MS, Killham K (1992) Factors affecting movement of microorganisms in soils. In: Stotzky G, Bolla LM (eds) Soil Biochemistry, 7, Marcel Dekker, New York. pp 263-305
98. Gandhi M, Sangwan V, Kapoor KK, Dilbaghi N (1997) Composting of household wastes with and without earthworms. Environ Ecol 15:432-434
99. Ganesh kumar A, Sekaran G (2005) Enteric pathogen modification by anaecic earthworm, Lampito Mauritii. J Appl Sci Environ Mgt 9:15-17
100. Gange AC (1993) Translocation of mycorrhizal fungi by earthworms during early succession. Soil Biol Biochem 25:1021-1026
101. Garcia C, Hernandez T, Costa F (1997) Potential use of dehydrogenase activity as an index of microbial activity in degraded soils. Commun Soil Sci Plant Anal 28:123-134
102. Garg P, Gupta A, Satya S (2006) Vermicomposting of different types of waste using Eisenia foetida: a comparative study. Bioresour Technol 97:391-395
103. Gavrilov K (1963) Earthworms, producers of biologically active substances. Zh Obshch Biol 24:149-154
104. Ghosh M, Chattopadhyay GN, Baral K (1999) Transformation of phosphorus during vermicomposting. Bioresour Technol 69:149-154
105. Gliotti C, Giusquiani PL, Businelli D, Machioni A (1997) Composition changes of dissolved organic matter in a soil amended with municipal waste compost. Soil Sci 162:919-926
106. Goenadi DH, Sudharama IM (1995) Shoot initiation by humic acids of selected tropical crops grown in tissue culture. Plant Cell Rep 15:59-62

107. Goldstein J (1998) Compost suppresses diseases in the lab and fields. BioCycle 39:62-64

108. Gopal M, Gupta A, Sunil E, Thomas VG (2009) Amplification of plant beneficial microbial communities during conversion of coconut leaf substrate to vermicompost by Eudrilus sp. Curr Microbiol 59:15-20

109. Graff O, Makeschin F (1980) Beeinlussung des Ertrags von Weidelgrass (Lolium muttiflorum) Ausscheidungen von Regenwurmen dreier verschiedener Arten. Pedobiologia 20:176-180

110. Grappelli A, Galli E, Tomati U (1987) Earthworm casting effect on Agaricus bisporus fructification. Agrochimica 2:457-462

111. Guerrero RD, Guerrero LA (2006) Response of eggplant (Solanum melongena) grown in plastic containers to vermicompost and chemical fertilizer. Asia Life Sciences 15:199-204

112. Gunadi B, Blount C, Edward CA (2002) The growth and fecundity of Eisenia foetida (Savigny) in cattle solids pre-composted for different periods. Pedobiologia 46:15-23

113. Gunathilagraj K, Ravignanam T (1996) Vermicomposting of sericulture wastes. Madras Agric J 83:455-457

114. Gupta AK, Pankaj PK, Upadhyava V (2008) Effect of vermicompost, farm yard manure, biofertilizer and chemical fertilizers (N, P, K) on growth, yield and quality of lady's finger (Abelmoschus esculentus). Pollution Research 27:65-68

115. Hahn H, Bopp M (1968) A cytokinin test with high specificity. Planta 83:115-118

116. Han J, Sun L, Dong X, Cai Z, Yang H, Wang Y, Song W (2005) Characterization of a novel plant growth-promoting bacteria strain Delftia tsuruhatensis HR4 both as a diazotroph and a potential biocontrol agent against various pathogens. Syst Appl Microbiol 28:66-76

117. Hand P, Hayes WA, Frankland JC, Satchell JE (1988) Vermicomposting of cow slurry. Pedobiologia 31:199-209

118. Haritha Devi S, Vijayalakshmi K, Pavana Jyotsna K, Shaheen SK, Jyothi K, Surekha Rani M (2009) Comparative assessment in enzyme activities and microbial populations during normal and vermicomposting. J Environ Biol 30:1013-1017

119. Hartenstein R, Neuhauser EF, Kaplan DL (1979) Reproductive potential of the earthworm Eisenia foetida. Oecologia 43:329-340

120. Hartenstein R (1983) Assimilation by earthworm Eisenia fetida. In: Satchell JE (ed) Earthworm ecology. From Darwin to vermiculture, Chapman and Hall, London. pp 297-308

121. Hashemimajd K, Kalbasi M, Golchin A, Shariatmadari H (2004) Comparison of vermicompost and composts as potting media for growth of tomatoes. J Plant Nutr 27:1107-1123

122. Hassen A, Belguith K, Jedidi N, Cherif A, Cherif M, Boudabous A (2001) Microbial characterization during composting of municipal solid waste. Bioresour Technol 80:217-225

123. Hawida S, Kapari L, Ossipov V, Ramtala MJ, Ruuhola T, Haukioja E (2007) Foliar phenolics are differently associated with Epirrita autmnata growth and immuno competence. J Chem Ecol 33:1013-1023

124. Herms DA (2002) Effects of fertilization on insect resistance of woody ornamental plants. Environ Entomol 31:923-933

125. Hoitink HA, Kuter GA (1986) Efects of composts in growth media on soil-bome pathogens. In: Chen Y, Avnimelech Y (eds) The role of organic matter in modern agriculture, Martinus Nijhoff Publishers, Dordrecht. pp 289-306

126. Hoitink HA, Stone AG, Han DY (1997) Suppression of plant diseases by compost. Hort Sci 32:184-187

127. Hoitink HA, Grebus ME (1997) Composts and Control of Plant Diseases. In: Hayes MHB, Wilson WS (eds) Humic Substances Peats and Sludges Health and Environmental Aspects, Royal Society of Chemistry, Cambridge. pp 359-366

128. Holtzclaw KM, Sposito G (1979) Analytical properties of the soluble metal-complexing fractions in sludge-soil mixtures. IV. Determination of carboxyl groups in fulvic acid. Soil Sci Soc Am J 43:318-323

129. Huelsman MF, Edwards CA, Lawrence JL, Clarke-Harris DO (2000) A study of the effect of soil nitrogen levels on the incidence of insect pests and predators in Jamaican sweet potato (Ipomoea batatus) and Callaloo (Amaranthus). Proc Brighton Pest Control Conference: Pests and Diseases 8D–13:895-900

130. Inbal E, Feldman M (1982) The response of a hormonal mutant of common wheat to bacteria of the Azospirillium. Israel J Bot 31:257-263

131. Ismail SA (1995) Earthworms in soil fertility management. In: Thampan PK (ed) Organic Agriculture, pp. 77-100

132. Ismail SA (1997) Vermicology: The biology of Earthworms. Orient Longman Limited, Chennai.

133. Jadhav AD, Talashilkar SC, Pawar AG (1997) Influence of the conjunctive use of FYM, vermicompost andurea on growth and nutrient uptake in rice. J Maharashtra Agric Univ 22:249-250

134. Jagnow G (1987) Inoculation of cereal crops and forage grasses with nitrogen-fixing rhizosphere bacteria: a possible cause of success and failure with regard to yield response - a review. Z Pflanzenernaehr Dueng Bodenkde 150:361-368

135. Jambhekar H (1992) Use of earthworm as a potential source of decompose organic wastes. Proc Nat Sem Org Fmg, Coimbatore. pp 52-53

136. Jannsson RK, Smilowitz Z (1986) Influence of nitrogen on population parameters of potato insects: abundance, population growth and within-plant distribution of the green peach aphid, Myzus persicae (Homoptera: Aphididae). Environ Entomol 15:49-55

137. Jha BK, Gandhi Pragash M, Cletus J, Raman G, Sakthivel N (2009) Simultaneous phosphate solubilization potential and antifungal activity of new fluorescent pseudomonad strains, Pseudomonas aeruginosa, P. plecoglossicida and P. mosselii. W J Microbiol Biotech 25:573-581

138. Jolly JM, Lappin-Scott HM, Anderson JM, Clegg CD (1993) Scanning electron microscopy of the gut microflora of two earthworms: Lumbricus terrestris and Octolasion cyaneum. Microbial Ecol 26:235-245

139. Kale RD, Bano K (1986) Field Trials with vermicompost (vee comp. E. 8. UAS) on organic fertilizers. In: Dass MC, Senapati BK, Mishra PC (eds) Proceedings of the national seminar on organic waste utilization, Sri Artatrana Ront, Burla. pp 151-157

140. Kale RD, Bano K (1988) Earthworm cultivation and culturing techniques for the production of vee COMP83E UAS. Mysore J Agric Sci 2:339-344

141. Kale RD, Mallesh BC, Bano K, Bagyaray DJ (1992) Influence of vermicompost application on the available macronutrients and selected microbial populations in paddy field. Soil Biol Biochem 24:1317-1320

142. Kale RD (1995) Vermicomposting has a bright scope. Indian Silk 34:6-9

143. Kalembasa D (1996) The influence of vermicomposts on yield and chemical composition of tomato. Zesz Probl Post Nauk Roln 437:249-252

144. Kannangowa T, Utkhede RS, Paul JW, Punja ZK (2000) Effect of mesophilic and thermophilic composts on suppression of Fusarium root and stem rot of greenhouse cucumber. Canad J Microbiol 46:1021-1022

145. Karmegam N, Alagermalai K, Daniel T (1999) Effect of vermicompost on the growth and yield of greengram (Phaseolus aureus Rob.). Trop Agric 76:143-146

146. Karmegam N, Daniel T (2000) Effect of biodigested slurry and vermicompost on the growth and yield of cowpea (Vigna unguiculata (L.). Environ Ecol 18:367-370

147. Karmegam N, Daniel T (2008) Effect of vermi-compost and chemical fertilizer on growth and yield of Hyacinth Bean (Lablab purpureas). Dynamic Soil, Dynamic Plant, Global Science Books 2:77-81

148. Karsten GR, Drake HL (1995) Comparative assessment of the aerobic and anaerobic microfloras of earthworm guts and forest soils. Appl Environ Microbiol 61:1039-1044

149. Karsten GR, Drake HL (1997) Denitrifying bacteria in the earthworm gastrointestinal tract and in vivo emission of nitrous oxide (N2O) by earthworms. Appl Environ Microbiol 63:1878-1882

150. Kerry B (1988) Fungal parasites of cyst nematodes. In: Edwards CA, Stinner BR, Stinner D, Rabatin S (eds) Biological Interactions in Soil, Elsevier, Amsterdam. pp 293-306

151. Khambata SR, Bhat JV (1953) Studies on a new oxalate-decomposing bacterium, Pseudomonas oxalaticus. J Bacteriol 66:505-507

152. Knuutinen J, Palm H, Hakala H, Haimi J, Huhta V, Salminen J (1990) Polychlorinated phenols and their metabolites in soil and earthworms of a saw mill environment. Chemosphere 20:609-623

153. Kolodziej M, Kostecka J (1994) Some qualitative features of the cucumbers and carrots cultivated on the vermicompost. Zeszyty Naukowe Akademii Rolniczej W Krakowie 292:89-94

154. Kooch Y, Jalilvand H (2008) Earthworm as ecosystem engineers and the most important detritivors in forst soils. Pak J Boil Sci 11:819-825

155. Kostecka J, Blazej JB, Kolodziej M (1996a) Investigations on application of vermicompost in potatoes farming in second year of experiment. Zeszyty Naukowe Akademii Rolniczej W Krakowie 310:69-77

156. Koul O (2008) Phytochemicals and insect control: an antifeedant approach. Crit Rev Plant Sci 27:1-24

157. Krishnamoorthy RV, Vajranabhiah SN (1986) Biological activity of earthworm casts: An assessment of plant growth promoter levels in casts. Proc Indian Acad Sci (Anim Sci) 95:341-35

158. Kristufek V, Ravasz K, Pizl V (1993) Actinomycete communities in earthworm guts and surrounding soil. Pedobiologia 37:379-384

159. Kurowska A, Gora J, Kalemba D (1990) Effects of plant phenols on insects. Pol Wiad Chem 44:399-409

160. Kuter GA, Nelson GB, Hoitink HA, Madden LV (1983) Fungal population in container media amended with composted hardwood bark suppressive and conductive to Rhizoctonia damping-off. Phytopathology 73:1450-1456

161. Lavelle E, Barois I, Martin A, Zaidi Z, Schaefer R (1989) Management of earthworm populations in agro-ecosystems: A possible way to maintain soil quality? In: Clarholm M, Bergstrom L (eds) Ecology of Arable Land: Perspectives and Challenges, Kluwer Academic Publishers, London. pp 109-122

162. Lavelle P, Martin A (1992) Small-scale and large-scale effects of endogeic earthworms on soil organic matter dynamics in soils of the humid tropics. Soil Biol Biochem 12:1491-1498

163. Lazarovits G, Tenuta M, Conn KL, Gullino ML, Katan J, Matta A (2000) Utilization of high nitrogen and swine manure amendments for control of soil-bome diseases: efficacy and mode of action. Acta Hortic 5:559-564

164. Lazcano C, Gomez-Brandon M, Dominguez J (2008) Comparison of the effectiveness of composting and vermicomposting for the biological stabilization of cattle manure. Chemosphere 72:1013-1019

165. Lee KE (1985) Earthworms: Their Ecology and Relationships with Soils and Land Use. Academic Press, Sydney.

166. Lee YS, Bartlett RJ (1976) Stimulation of plant growth by humic substances. Soil Sci Soc Am J 40:876-879

167. Lunt HA, Jacobson HGM (1944) The chemical composition of earthworm casts. Soil Sci 58:367-375

168. Maboeta MS, Van Rensburg L (2003) Vermicomposting of industrially produced wood chips and sewage sludge utilizing Eisenia foetida. Ecotoxicol Environ Saf 56:265-270

169. Madsen EL, Alexander M (1982) Transport of Rhizobium and Pseudomonas through soil. Soil Sci Soc Am J 46:557-560

170. Mahanil S, Attajarusit J, Stout MJ, Thipayong P (2008) Over expression of tomato phenol oxidase increases resistance to the common cutworm. Plant Sci 174:456-466

171. Maheswarappa HP, Nanjappa HV, Hegde MR (1999) Influence of organic manures on yield of arrowroot, soil physico-chemical and biological properties when grown as intercrop in coconut garden. Ann Agr Res 20:318-323

172. Mahmoud SA, Ramadan Z, Thabet EM, Khater T (1984) Production of plant growth promoting substance rhizosphere organisms. Zentrbl Mikrobiol 139:227-232

173. Makulec G (2002) The role of Lumbricus rubellus Hoffm. In determining biotic and abiotic properties of peat soils. Pol J Ecol 50:301-339

174. Marinari S, Masciandaro G, Ceccanti B, Grego S (2000) Influence of organic and mineral fertilisers on soil biological and physical properties. Bioresour Technol 72:9-17

175. Martin JP (1976) Darwin on earthworms: the formation of vegetable moulds. Bookworm Publishing, Ontario.

176. Masciandaro G, Ceccanti B, Gracia C (1997) Soil agro-ecological management: fertigation and vermicompost treatments. Bioresour Technol 59:199-206

177. Masciandaro G, Ceccanti B, Gracia C (2000) `In situ' vermicomposting of biological sludges and impacts on soil quality. Soil Biol Biochem 32:1015-1024

178. Meena RN, Singh Y, Singh SP, Singh JP, Singh K (2007) Effect of sources and level of organic manure on yield, quality and economics of garden pea (Pisum sativam L.) in eastern uttar pradesh. Vegetable Science 34:60-63

179. Mitchell A, Edwards CA (1997) The production of vermicompost using Eisenia fetida from cattle manure. Soil Biol Biochem 29:3-4

180. Mitchell A (1997) Production of Eisenia fetida and vermicompost from feed-lot cattle manure. Soil Biol Biochem 29:763-766

181. Mitchell MJ, Hornor SG, Abrams BI (1980) Decomposition of sewage sludge in drying beds and the potential role of the earthworm, Eisenia foetida. J Environ Qual 9:373-378

182. Mitchell MJ (1978) Role of invertebrates and microorganisms in sludge deomposition. In: Hartenstein R (ed) Utilisation of soil organisms in sludge management. natural technology information services, Springfield, Virginia. pp 35-50

183. Monroy F, Aira M, Domínguez J (2009) Reduction of total coliform numbers during vermicomposting is caused by short-term direct effects of earthworms on microorganisms and depends on the dose of application of pig slurry. Sci Tot Environ 407:5411-5416

184. Moody SA, Piearce TG, Dighton J (1996) Fate of some fungal spores associated with wheat straw decomposition on passage through the guts of Lumbricus terrestris and Aporrectodea longa. Soil Biol Biochem 28:533-537

185. Morra L, Palumbo AD, Bilotto M, Ovieno P, Ptcascia S (1998) Soil solarization: organic fertilization grafting contribute to build an integrated production system in a tomato-zucchini sequence. Colture-Protte 27:63-70

186. Munnoli PM, Da Silva JAT, Saroj B (2010) Dynamics of the soil-earthworm-plant relationship: a review. Dynamic soil, dynamic plant. 1-21

187. Muscolo A, Bovalo F, Gionfriddo F, Nardi S (1999) Earthworm humic matter produces auxin-like effect on Daucus carota cell growth and nitrate metabolism. Soil Biol Biochem 31:1303-1311

188. Muscolo A, Felici M, Concheri G, Nardi S (1993) Effect of earthworm humic substances on esterase and peroxidase activity during growth of leaf explants of Nicotiana plumbaginifolia. Biol Fertil Soils 15:127-131

189. Muscolo A, Panuccio MR, Abenavoli MR, Concheri G, Nardi S (1996) Effect of molecular complexity acidity of earthworm faeces humic fractions on glutamale dehydrogenase, glutamine synthetase, and phosphoenolpyruvate carboxylase in Daucus carota II cells. Biol Fertil Soils 22:83-88

190. Mylonas VA, Mccants CB (1980) Effects of humic and fulvic acids on growth of tobacco. I. Root initiation and elongation. Plant Soil 54:485-490

191. Nagavallemma KP, Wani SP, Stephane L, Padmaja VV, Vineela C, Babu Rao M, Sahrawat KL (2004) Vermicomposting: Recycling wastes into valuable organic fertilizer. Global Theme on Agrecosystems Report no.8. Patancheru 502324. International Crops Research Institute for the Semi-Arid Tropics, Andhra Pradesh. p 20

192. Nakamura Y (1996) Interactions between earthworms and microorganisms in biological control of plant root pathogens. Farming Jpn 30:37-43

193. Nakasone AK, Bettiol W, de Souza RM (1999) The effect of water extracts of organic matter on plant pathogens. Summa Phytopathology 25:330-335

194. Nardi S, Arnoldi G, Dell'Agnola G (1988) Release of hormone-like activities from Alloborophora rosea and Alloborophora caliginosa feces. J Soil Sci 68:563-657

195. Nardi S, Dell'Agnola G, Nuti PM (1983) Humus production from farmyard wastes by vermicomposting. Proc. Int. Symp. On Agricultural and Environmental Prospects in Earthworm Farming. Rome. pp 87-94

196. Ndegwa PM, Thompson SA, Das KC (2000) Effects of stocking density and feeding rate on vermicomposting of biosolids. Bioresour Technol 71:5-12

197. Nechitaylo TY, Yakimov MM, Godinho M, Timmis KN, Belogolova E, Byzov BA, et al. (2010) Effect of the earthworms Lumbricus terrestris and Aporrectodea caliginosa on bacterial diversity in soil. Microbial Ecol 59:574-587

198. Nethra NN, Jayaprasad KV, Kale RD (1999) China aster (Callistephus chinensis (L)) cultivation using vermicompost as organic amendment. Crop Research, Hisar 17:209-215

199. Nielson RL (1965) Presence of plant growth substances in earthworms demonstrated by paper chromatography and the Went pea test. Nature 208:1113-1114

200. Orozco FH, Cegarra J, Trujillo LM, Roig A (1996) Vermicomposting of coffee pulp using the earthworm Eisenia fetida: effects on C and N contents and the availability of nutrients. Biol Fertil Soils 22:162-166

201. Park SR, Cho EJ, Yu KH, Kim YS, Suh JJ, Chang CS (1996) Endogenous phenoloxidase from an earthworm Lumbricus rubellus. Tongmul Hakoehi 39:36-46

202. Parle JN (1963) A Microbiological Study of Earthworm Casts. J Gen Microbiol 31:13-22

203. Parthasarathi K, Ranganathan LS (1998) Pressmud vermicast are hot spots of fungi and bacteria. Ecol Environ Cons 4:81-86

204. Pathma J, Ayyadurai N, Sakthivel N (2010) Assessment of Genetic and Functional Relationship of Antagonistic Fluorescent Pseudomonads of Rice Rhizosphere by Repetitive Sequence, Protein Coding Sequence and Functional Gene Analyses. J Microbiol 48:715-727

205. Pathma J, Kamaraj Kennedy R, Sakthivel N (2011a) Mechanisms of fluorescent pseudomonads that mediate biological control of phytopathogens and plant growth promotion of crop plants. In: Maheswari DK (ed) Bacteria in Agrobiology: Plant Growth Responses, SpringerVerlag, Berlin. pp 77-105

206. Pathma J, Rahul GR, Kamaraj Kennedy R, Subashri R, Sakthivel N (2011) Secondary metabolite production by bacterial antagonists. Journal of Biological Control 25:165-181

207. Patil SL, Sheelavantar MN (2000) Effect of moisture conservation practices, organic sources and nitrogen levels on yield, water use and root development of rabi sorghum (Sorghum bicolor (L.)) in the vertisols of semiarid tropics. Ann Agric Res 21:32-36

208. Patriquin DG, Baines D, Abboud A (1995) Diseases, pests and soil fertility. In: Cook HF, Lee HC (eds) Soil Management in Sustainable Agriculture, Wye College Press, Wye. pp 161-174

209. Pedersen JC, Hendriksen NB (1993) Effect of passage through the intestinal tract of detritivore earthworms (Lumbricus spp.) on the number of selected gram-negative and total bacteria. Biol Fertil Soils 16:227-232

210. Petersen H, Luxton MA (1982) A comparative analysis of soil fauna populations and their role in decomposition process. Oikos 39:287-388

211. Phelan PL, Norris KH, Mason JF (1996) Soil management history and host preference by Ostrinia nubilatis: evidence for plant mineral balance mediating insect-plant interactions. Environ Entom 25:1329-1336

212. Phelan PL (2004) Connecting below-ground and above-ground food webs: the role of organic matter in biological buffering. In: Magadoff F, Well RR (eds) Soil Organic Matter in sustainable agriculture, CRC Press, Boca Raton. pp 199-226

213. Pinel N, Davidson SK, Stahl DA (2008) Verminephrobacter eiseniae gen. nov., sp. nov., a nephridial symbiont of the earthworm Eisenia foetida (Savigny). Int J Syst Evol Microbiol 58:2147-2157

214. Pitt D, Tilston EL, Groenhof AC, Szmidt RA (1998) Recycled organic materials (ROM) in the control of plant disease. Acta Hortic 469:391-403

215. Pizl V, Novokova A (1993) Interactions between microfungi and Eisenia andrei (Oligochaeta) during cattle manure vermicomposting. Pedobiologia 47:895-899

216. QiTian S (2004) Research on prevention and elimination of agricultural pests by ginkgo phenols phenolic acids. Chem Ind For Prod 24:83

217. Raguchander T, Rajappan K, Samiyappan R (1998) Influence of biocontrol agents and organic amendments on soybean root rot. Int J Trop Agri 16:247-252

218. Ramesh P (2000) Effects of vermicomposts and vermicornposting on damage by sucking pests to ground nut (Arachis hypogea). Indian J Agri Sci 70:334

219. Rao KR, Rao PA, Rao KT (2001) Influence of fertilizers and manures on the population of coccinellid beetles and spiders in groundnut ecosystem. Ann Plant Protect Sci 9:43-46

220. Rao KR (2002) Induce host plant resistance in the management sucking pests of groundnut. Ann Plant Protect Sci 10:45-50

221. Rao KR (2003) Influence of host plant nutrition on the incidence of Spodoptera litura and Helicoverpa armigera on groundnuts. Indian J Entomol 65:386-392

222. Ravindra NP, Raman G, Badri Narayanan K, Sakthivel N (2008) Assessment of genetic and functional diversity of phosphate solubilizing fluorescent pseudomonads isolated from rhizospheric soil. BMC Microbiol 8:230

223. Reeh U (1992) Influence of population densities on growth and reproduction of the earthworm Eisenia andrei on pig manure. Soil Biol Biochem 24:1327-1331

224. Ribeiro CF, Mizobutsi EH, Silva DG, Pereira JCR, Zambolim L (1998) Control of Meloidognye javanica on lettuce with organic amendments. Fitopatol Brasileira 23:42-44

225. Riffaldi R, Levi-Minzi R (1983) Osservazioni preliminari sul ruolo dell Eisenia foetida nell'umificazione del letame. Agrochimica 27:271-274

226. Rivera AMC, Wright ER, López MV, Fabrizio MC (2004) Temperature and dosage dependent suppression of damping-off caused by Rhizoctonia solani in vermicompost amended nurseries of white pumpkin. Phyton 53:131-136

227. Rodriguez JA, Zavaleta E, Sanchez P, Gonzalez H (2000) The effect of vermicomposts on plant nutrition, yield and incidence of root and crown rot of gerbera (Gerbera jamesonii H. Bolus). Fitopatol 35:66-79

228. Rodriguez-Kabana R (1986) Organic and inorganic amendments to soil as nematode suppressants. J Nematol 18:129-135

229. Rouelle J (1983) Introduction of an amoeba and Rhizobium Japonicum into the gut of Eisenia fetida (Sav.) and Lumbricus terrestris L. In: Satchell JE (ed) Earthworm Ecology: From Darwin to Vermiculture, Chapman and Hall, New York. pp 375-381

230. Sánchez-Monedero MA, Roig A, Paredes C, Bernal MP (2001) Nitrogen transformation during organic waste composting by the Rutgers system and its effects on ph, EC and maturity of the composting mixtures. Bioresour Technol 78:301-308

231. Sainz MJ, Taboada-Castro MT, Vilariño A (1998) Growth, mineral nutrition and mycorrhizal colonization of red clover and cucumber plants grown in a soil amended with composted urban wastes. Plant Soil 205:85-92

232. Saumaya G, Giraddi RS, Patil RH (2007) Utility of vermiwash for the management of thrips and mites on chilli (Capiscum annum) amended with soil organics. Karnataka J Agric Sci 20:657-659

233. Scheu S (1992) Automated measurement of the respiratory response of soil microcompartments: active microbial biomass in earthworm faeces. Soil Biol Biochem 24:1113-1118

234. Scheuerell SJ, Sullivan DM, Mahaffee WF (2005) Suppression of seedling damping-off caused by Pythium ultimum, and Rhizoctonia solani in container media amended with a diverse range of Pacific Northwest compost sources. Phytopathology 95:306-315

235. Schmidt O, Doubre BM, Ryder MH, Killman K (1997) Population dynamics of Pseudomonas corrugata 2140R LUX8 in earthworm food and in earthworm cast. Soil Biol Biochem 29:523-528

236. Sembdner G, Borgman E, Schneider G, Liebisch HW, Miersch O, Adam G, Lischewski M, Schieber K (1976) Biological activity of some conjugated gibberellins. Planta 132:249-257

237. Senesi N, Saiz-Jimenez C, Miano TM (1992) Spectroscopic characterization of metal-humic acid-like complexes of earthworm-composted organic wastes. Sci Total Environ 117–118:111-120

238. Sharma S, Pradhan K, Satya S, Vasudevan P (2005) Potentiality of earthworms for waste management and in other uses - A Review. The Journal of American Science 1:4-16

239. Sharpley AN, Syers JK (1976) Potential role of earthworm casts for the phosphorous enrichment of runoff waters. Soil Biol Biochem 8:341-346

240. Shi-wei Z, Fu-zhen H (1991) The nitrogen uptake efficiency from 15N labeled chemical fertilizer in the presence of earthworm manure (cast). In: Veeresh GK, Rajagopal D, Viraktamath CA (eds) Advances in Management and Conservation of Soil Fauna, Oxford and IBH publishing Co, New Delhi. pp 539-542

241. Siddiqui ZA, Mahmood I (1999) Role of bacteria in the management of plant parasitic nematodes: a review. Bioresour Technol 69:167-179

242. Sidhu J, Gibbs RA, Ho GE, Unkovich I (2001) The role of indigenous microorganisms in suppression of Salmonella regrowth in composted biosolids. Water Res 35:913-920

243. Simsek Ersahin Y, Haktanir K, Yanar Y (2009) Vermicompost suppresses Rhizoctonia solani Kühn in cucumber seedlings. J Plant Dis Protect 9:15-17

244. Singh R, Sharma RR, Kumar S, Gupta RK, Patil RT (2008) Vermicompost substitution influences growth, physiological disorders, fruit yield and quality of strawberry (Fragaria x ananassa Duch.). Bioresour Technol 99:8507-8511

245. Singh UP, Maurya S, Singh DP (2003) Antifungal activity and induced resistance in pea by aqueous extract of vermicompost and for control of powdery mildew of pea and balsam. J Plant Dis Protect 110:544-553

246. Singleton DR, Hendrixb PF, Colemanb DC, Whitmana WB (2003) Identification of uncultured bacteria tightly associated with the intestine of the earthworm Lumbricus rubellus (Lumbricidae; Oligochaeta). Soil Biol Biochem 35:1547-1555

247. Sinha RK, Agarwal S, Chauhan K, Valani D (2010) The wonders of earthworms and its vermicompost in farm production: Charles Darwin's 'friends of farmers', with potential to replace destructive chemical fertilizers from agriculture. Agricultural sciences 1:76-94

248. Sinha RK, Bharambe G, Chaudhari U (2008) Sewage treatment by vermifiltration with synchronous treatment of sludge by earthworms: a low-cost sustainable technology over conventional systems with potential for decentralization. The Environmentalist 28:409-420

249. Sinha RK, Heart S, Agarwal S, Asadi R, Carretero E (2002) Vermiculture technology for environmental management: study of the action of the earthworms Eisenia foetida, Eudrilus euginae and Perionyx excavatus on biodegradation of some community wastes in India and Australia. The Environmentalist 22:261-268

250. Sinha RK, Herat S, Valani D, Chauhan K (2009) Vermiculture and sustainable agriculture. Am-Euras J Agric and Environ Sci, IDOSI Publication 5:1-55

251. Sipes BS, Arakaki AS, Schmitt DP, Hamasaki RT (1999) Root-knot nematode management in tropical cropping systems with organic products. J Sustain Agr 15:69-76

252. Sreenivas C, Muralidhar S, Rao MS (2000) Vermicompost, a viable component of IPNSS in nitrogen nutrition of ridge gourd. Ann Agr Res 21:108-113

253. Steffen KL, Dan MS, Harper JK, Fleischer SJ, Mkhize SS, Grenoble DW, MacNab AA, Fager K (1995) Evaluation of the initial season for implementation of four tomato production systems. J Am Soc Hort Sci 120:148-156

254. Stephens PM, Davoren CW, Doube BM, Ryder MH (1993) Reduced superiority of Rhizoctonia solani disease on wheat seedlings associated with the presence of the earthworm Aporrectodea trapezoids. Soil Biol Biochem 11:1477-1484

255. Stephens PM, Davoren CW, Ryder MH, Doube BM, Correll RL (1994) Field evidence for reduced severity of Rhizoctonia bare-patch disease of wheat, due to the presence of the earthworms Aporrectodea rosea and Aporrectodea trapezoides. Soil Biol Biochem 26:1495-1500

256. Stephens PM, Davoren CW, Ryder MH, Doube BM (1994) Influence of the earthworm Aporrectodea trapezoides (Lumbricidae) on the colonization of alfalfa (Medicago sativa L.) roots by Rhizobium melilotti strain LS-30R and the survival of L5-30R in soil. Biol Fertil Soils 18:63-70

257. Stephens PM, Davoren CW (1997) Influence of the earthworms Aporrectodea trapezoides and A. rosea on the disease severity of Rhizoctonia solani on subterranean cloves and ryegrass. Soil Biol Biochem 29:511-516

258. Stone AG, Scheurell SJ, Darby HM (2004) Suppression of soilborne diseases in field agricultural systems: organic matter management, cover cropping and other cultural practices. In: Magdoff F, Weil (eds) Soil Organic Matter in Sustainable Agriculture, CRC Press LLC, Boca Raton. pp 131-177

259. Subler S, Edwards CA, Metzger PJ (1998) Comparing vermicomposts and composts. Biocycle 39:63-66

260. Sudhakar K, Punnaiah KC, Krishnayya PV (1998) Influence of organic and inorganic fertilizers and certain insecticides on the incidence of shoot and fruit borer, Leucinodes orbonalis Guen, infesting brinjal. J Entomol Res 22:283-286

261. Suhane RK (2007) Vermicompost. Publication of Rajendra Agriculture University, Pusa. 88

262. Summers G, Felton GW (1994) Prooxidation effects of phenolic acids on the generalist herbivore Helicoverpa zea: potential mode of action of phenolic compounds on plant anti-herbivory chemistry. Insect Biochem Mol Biol 24:943-953

263. Sunish KR, Ayyadurai N, Pandiaraja P, Reddy AV, Venkateshwarlu Y, Prakash O, Sakthivel N (2005) Characterization of antifungal metabolite produced by a new strain Pseudomonas aeruginosa PuPa3 that exhibits broad-spectrum antifungal activity and biofertility traits. J Appl Microbiol 98:145-154

264. Suthar S (2010) Evidence of plant hormone like sub-stances in vermiwash: An ecologically safe option of synthetic chemicals for sustainable farming. J Ecol Eng 36:1089-1092

265. Suthar S, Singh S (2008) Vermicomposting of domestic waste by using two epigeic earthworms (Perionyx excavatus and Perionyx sansibaricus). Int J Evniron Sci and Technol 5:99-106

266. Swathi P, Rao KT, Rao PA (1998) Studies on control of root-knot nematode Meloidogyne incognita in tobacco miniseries. Tobacco Res 1:26-30

267. Szcech M, Rondomanski W, Brzeski MW, Smolinska U, Kotowski JF (1993) Suppressive effect of a commercial earthworm compost on some root infecting pathogens of cabbage and tomato. Biol Agric and Hortic 10:47-52

268. Szczech M, Smolinska U (2001) Comparison of suppressiveness of vermicomposts produced from animal manures and sewage sludge against Phytophthora nicotianae Breda de Haan var. nicotiannae. J Phytopathology 149:77-82

269. Szczech MM (1999) Suppressiveness of vermicomposts against fusarium wilt of tomato. J Phytopathology 147:155-161

270. Tajbakhsh J, Abdoli MA, Mohammadi Goltapeh E, Alahdadi I, Malakouti MJ (2008) Trend of physico-chemical properties change in recycling spent mushroom compost through vermicomposting by epigeic earthworms Eisenia foetida and E. andrei. J Agric Technol 4:185-198

271. Tan KH, Tantiwiramanond D (1983) Effect of humic acids on nodulation and dry matter production of soybean, peanut, and clover. Soil Sci Soc Am J 47:1121-1124

272. Thoden TC, Korthals GW, Termorshuizen (2011) Organic amendments and their influences on plant-parasitic and free living nematodes: a promosing method for nematode management. Nematology 13:133-153

273. Tiquia SM (2005) Microbiological parameters as indicators of compost maturity. J Appl Microbiol 99:816-828

274. Tiunov AV, Scheu S (2000) Microfungal communities in soil litter and casts of Lumbricus terrestris (Lumbricidae): a laboratory experiment. Appl Soil Ecol 14:17-26

275. Tiwari SC, Tiwari BK, Mishra RR (1989) Microbial populations, enzyme activities and nitrogen, phosphorous, potassium enrichment in earthworm casts and in the surrounding soil of pine apple plantation. Biol Fertil Soils 8:178-182

276. Tomati U, Grapppelli A, Galli E (1987) The presence of growth regulators in earthworm worked waste. In: Bonvicini Paglioi AM, Omodeo P (eds) On Earthworms. Proceedings of International Symposium on Earthworms, Selected Symposia and Monographs, Union Zoologica Italian, 2, Modena, Mucchi. pp 423-435

277. Tomati U, Grapppelli A, Galli E (1988) The hormone-like effect of earthworm casts on plant growth. Biol Fertil Soils 5:288-294

278. Toyota K, Kimura M (2000) Microbial community indigenous to the earthworm Eisenia foetida. Biol Fertil Soils 31:187-190

279. Trevors JT (1984) Dehydrogenase activity in soil. A comparison between the INT and TTC assay. Soil Biol Biochem 16:673-674

280. Umesh B, Mathur LK, Verma JN, Srivastava (2006) Effects of vermicomposting on microbiological flora of infected biomedical waste. ISHWM Journal 5:28-33

281. Vadiraj BA, Siddagangaiah D, Potty SN (1998) Response of coriander (Coriandrum sativum L.) cultivars to graded levels of vermicompost. J Spices Aromatic Crops 7:141-143

282. Valdrighi MM, Pera A, Agnolucci M, Frassinetti S, Lunardi D, Vallini G (1996) Effects of compost-derived humic acids on vegetable biomass production and microbial growth within a plant (Cichorium intybus) soil system: a comparative study. Agric Ecosyst Environ 58:133-144

283. Valenzuela O, Gluadia Y, Gallardo S (1997) Use of vermicompost as a growing medium for tomato seedlings (cv. Pltense). Revista Cientifica Agropecuaria 1:15-21

284. Vaz-Moreira I, Maria E, Silva CM, Manaia Olga C, Nunes (2008) Diversity of Bacterial Isolates from Commercial and Homemade Composts. Microbial Ecol 55:714-722

285. Vinken R, Schaeffer A, Ji R (2005) Abiotic association of soil-borne monomeric phenols with humic acids. Org Geochem 36:583-593

286. Vivas A, Moreno B, Garcia-Rodriguez S, Benitez E (2009) Assessing the impact of composting and vermicomposting on bacterial community size and structure, and functional diversity of an olive-mill waste. Bioresour Technol 100:1319-1326

287. Webster KA (2005) Vermicompost increases yield of cherries for three years after a single application. EcoResearch, South Australia.

288. Weltzien HC (1989) Some effects of composted organic materials on plant health. Agric Ecosyst Environ 27:439-446

289. Wilson DP, Carlile WR (1989) Plant growth in potting media containing worm-worked duck waste. Acta Hortic 238:205-220

290. Yardim EN, Arancon NQ, Edwards CA, OliverTJ BRJ (2006) Suppression of tomato hornworm (Manduca quinquemaculata) and cucumber beetles (Acalymma vittatum and Diabotrica undecimpunctata) populations and damage by vermicomposts. Pedobiologia 50:23-29

291. Yardim EN, Edwards CA (2003) Effects of organic and synthetic fertilizer sources on pest and predatory insects associated with tomatoes. Phytoparasitica 31:324-329

292. Yasir M, Aslam Z, Kim SW, Lee SW, Jeon CO, Chung YR (2009) Bacterial community composition and chitinase gene diversity of vermicompost with antifungal activity. Bioresour Technol 100:4396-4403

293. Yasir M, Aslam Z, Song GC, Jeon CO, Chung YR (2009b) Eiseniicola composti gen. nov., sp. nov., with antifungal activity against plant pathogenic fungi. Int J Sys Evol Microbiol 60:268

294. Yeates GW (1981) Soil nematode populations depressed in the presence of earthworms. Pedobiologiaogia 22:191-202

295. Zhang BG, Li GT, Shen TS, Wang JK, Sun Z (2000) Changes in microbial biomass C, N, and P and enzyme activities in soil incubated with the earthworms Metaphire guillelmi or Eisenia foetida. Soil Biol Biochem 32:2055-2062

PART III

BIODRYING

CHAPTER 9

Criteria for Assessing the Viablility of a Small Scale MSW Bio-Drying Plant Aimed to RDF Production for Local Use

E. C. RADA AND M. RAGAZZI

9.1 INTRODUCTION

One of the leading principles for EU countries is the concept of a waste management where in the first place there is the waste prevention, in the second one the most desirable options, recycling and energy generation, and in the last one the disposal of waste with no recovery of either materials and/or energy. The technologies who attracted considerable interest in the latest years are the Mechanical Biological Treatment (MBT) and the Biological Mechanical Treatment (BMT) (Muller, 2009; Tintner et al., 2009; Ragazzi and Rada 2009a). A part of the output from these processes, named generally Refuse Derived Fuel (RDF), can have industrial uses, for instance as a substitute of coal (Ragazzi and Rada, 2009b; Rada et al., 2005b; Gleis 2009).

Rada EC and Ragazzi M. "Criteria for Assessing the Viablility of a Small Scale MSW Bio-Drying Plant Aimed to RDF Production for Local Use." SARDINA 2009, Twelfth International Waste Management and Landfill Symposium, S. Margherita di Pula, Cagliari, Italy; 5 - 9 October 2009. CISA Publisher (2009). Used with permission from the publisher.

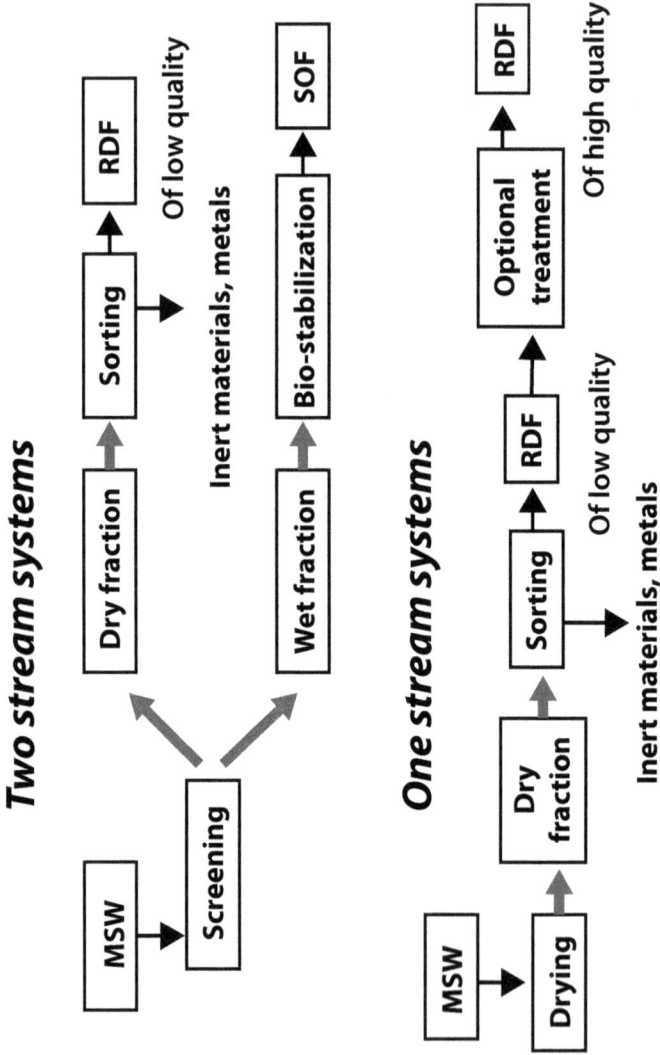

FIGURE 1: Two concepts of MBT plants

In the frame of Municipal Solid Waste (MSW) management, one of the topics more discussed is the treatment of the residual fraction that must be handled even when a good selective collection is performed. This fraction is named residual MSW (RMSW) and, from the theoretical point of view, should be composed exclusively by non-recyclable materials. In real scale the separation at the source cannot be perfect, thus RMSW has a variable composition, but generally one of the available options for its handling can be based on the generation of RDF. To this concern a typical strategy is co-combustion in cement works. In this case the partial substitution of coal follows the principles of the Kyoto protocol as a part of RDF is biomass.

Among the options for RDF generation, the two schemes that are often adopted are presented in Figure 1. One is based on the concept of two-streams treatment of MSW: an initial sieving separates MSW into an oversieve aimed to RDF generation and into an undersieve aimed to a biological treatment. The undersieve is generally too contaminated for allowing a generation of a good compost. This is the critical point of this strategy.

In the field of aerobic processes a distinction must be made between bio-stabilization and bio-drying (Rada et al., 2006a):

- Bio-stabilization: treatment of aerobic bioconversion generally applied to the fine fraction of residual MSW (under-sieve from mechanical selection, etc.) with the aim of generating a stabilized organic fraction (SOF) suitable to be landfilled or used for capping operations during the cultivation of the sanitary landfill;
- Bio-drying: treatment of aerobic bioconversion applied to residual MSW. After an inert separation, the final product is aimed to the use as RDF.

In the last two decades another approach has been developed: the one-stream treatment. In this case the main initial stage is the biological one, where a bio-drying process is performed. All the waste entering into the plant is introduced into the aerobic reactor with the target of decreasing the MSW water content, thanks to the exothermal reactions of a biochemical oxidation. The bio-dried material is easily refined in order to extract recyclable materials (glass, metals and inert). The obtained product after refinement could be classified as RDF, but in some cases, a post-sieving is required in order to separate the fine fractions (with a low energy content) and to obtain a high quality RDF (RDF_HQ). This post-sieving seems to be compulsory when a very high Lower Heating Value (LHV) must be guaranteed for the generated RDF.

Bio-drying is generally proposed for a decentralized management of MSW: a few satellite plants could generate RDF for a centralized use. This decentralization could be based on a small scale plant for local RDF exploitation. To this concern in the MSW sector there is not yet an approach useful for assessing the viability of bio-drying when a factory requiring heat wants to generate exactly the amount of RDF to be used in its own dedicated RDF burner.

In the present paper some criteria for assessing viability of this strategy (bio-drying plant coupled with a strictly dedicated RDF burner) are presented.

9.2 METHOD

9.2.1 DEFINITION OF THE CRITERIA

The criteria to be used for the analysis of the viability of bio-drying at small scale for industrial uses aimed to heat generation can be related to the following concepts:

- heat needed from the factory;
- costs for the conventional fuel to be substituted;
- residual MSW characteristics in the area around the factory;
- expected bio-drying performances referred to the available residual MSW in the area;
- assessment of RDF characteristics after post-refining;
- assessment of the amount of RDF needed for generating the required heat;
- assessment of MSW needed as input of the system to be implemented;
- assessment of the population involved for guaranteeing the MSW availability;
- assessment of the capital cost for MSW treatment;
- assessment of operating costs for MSW treatment;
- assessment of the capital cost for the RDF burner serviced to the factory;
- assessment of operating costs for the RDF burner;
- analysis of thermally integrated solutions (in case of variable energy demand);

- analysis of waste integrated solutions (in case of availability of special waste too);
- economic comparison with conventional fuels adoption.

9.2.2   CRITICAL ASPECTS OF THE CRITERIA

9.2.2.1 HEAT NEEDED FROM THE FACTORY

The method presented in this paper concerns the small scale applications of RDF, thus the target is basically related to heat generation (electricity generation needs a larger scale as the efficiency of the most conventional option for co-generation, that is a steam cycle, is very low at small scale). As a consequence it is important to know the amount of MWh_{th} y^{-1} necessary for the plant where the substitution of conventional fuels by RDF must be analysed. It is clear that the matter is not only the stream of heat on yealy basis as, looking at a deeper detail this option, the characteristics of the heat generated by RDF must be compatible with the characteristics of the heat necessary for the plant. In this paper we suppose that the need of heat can be satisfied by the generation of off-gas (from RDF burning) at high temperature.

9.2.2.2 COSTS FOR THE CONVENTIONAL FUEL

When some calculations must be performed with long-time scenarios a big trouble is related to the cost of conventional fuel in a period that is even longer than 10 years. Suppose to decide to construct a RDF based strategy. The timing for the implementation of the system could be 2 years (indeed a preliminary step for desining and authorisation must be taken into account). The lasting of the RDF plants for the new strategy will be 15 years. Thus the cost-benefits analysis must take into account the dynamics of costs for conventional fuels for the next 17 years. Concerning the trend for conventional fuels it is clear that this parameter is not steady and can fluctuate significantly in the short time. The trend in the long time, more related to the calculations interesting for this paper, can be anyway assumed growing.

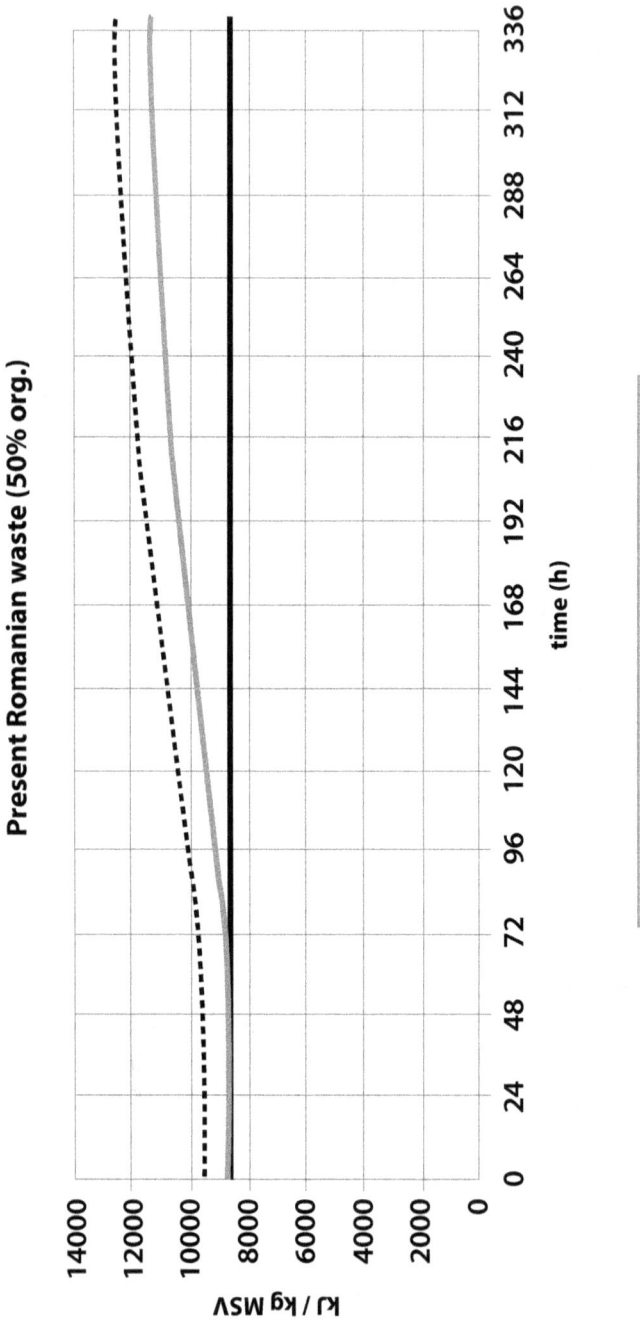

FIGURE 2: LHV dynamics during the bio-drying process.

9.2.2.3 RESIDUAL MSW CHARACTERISTICS

This is another critical aspect of the calculations as the MSW management can change dramatically if there is not a clear plan in the region where the plant is located. In particular, the strategy and the efficiency of selective collection can strongly modify the characteristics and the amount of MSW available in the area. More in general the economic development can increase the amount of MSW generated on per-capita base and can increase the energy content of MSW because of the increase of packaging. In the same time, the activation of more and more efficient strategies of selective collection can decrease the amount of MSW available for the studied strategy. In order to limit this problem it is important to set a strong agreement with the local authorities responsible for the management of MSW. Particular attention must be paied if the bio-drying process is proposed in an area where a high efficiency of organic fraction selective collection is performed or planned: in this case the organic fraction in the residual MSW can drop down to values that are not compatible with bio-drying (Rada et al., 2009a).

9.2.2.4 BIO-DRYING PERFORMANCES

An example of the Lower Heating Value (LHV) dynamics of RMSW, bio-dried material and RDF during the bio-drying treatment is reported in Figure 2, (Rada et al., 2008). In the reported example (referred to an area of Romania) the biodried material and LHV increase of RDF after two weeks is respectively around 30% and 44%. This is not an energy increase because it must be taken into account that the available mass of fuel is lower after the process.

The process allows "concentrating" the initial energy with a contemporary consumption of electrical energy. Generally the energy available at the end of the process is about 3% lower than the initial one (Rada 2005a), apart from the electricity needs that changes depending on the adopted technology. On large scale this two aspects are a weakness of the strategy if the final target were a burning a bio-dried material: you would pay for having less energy, but if the pre-treatment of residul MSW is aimed to the generation of a "product" (RDF) that open towards new options, this contradiction is no longer present. The international literature (Rada et al., 2006b; Rada et al., 2007a,b; Shao et al., 2009; Scotti 2009; Rada et al., 2009b; Grosso et al., 2009) make it available all the parameters needed to assess the behaviour of bio-drying referred to a residual MSW with a known composition.

9.2.2.5 RDF CHARACTERISTICS

From the theoretical point of view, a first calculation of the characteristics of RDF can be made assuming RDF as a refined bio-dried material: a simplified refining can be performed by separating metals, glass and inert. Depending on the initial value of LHV for the treated waste, the result of this simplified separation could be an RDF with a LHV higher or lower than 15 MJ kg^{-1} that could be assumed as a threshold for a viable result if RDF has to be used in a simplified burner (Rada 2005a). In case there is a need of increasing the LHV of RDF, an additional post-refining is necessary but a clear consequence is an increase in the amount of residues to be landfilled.

9.2.2.6 RDF AMOUNT

The amount of RDF needed is related to the amount of heat to be guaranteed at the factory. There are some details to be taken into account: the generation of RDF is not 365 days per year as at least a couple of weeks must be planned for maintenance problems. Also the use of RDF is not performed 365 days per year as the industrial process of most of the factories is based on a stop of at least a couple of weeks per year. This last stop could be synchronised with the first stop, but a correct design of the system should take into account also unfavourable situations. As a consequence there is the need of a (small) storage area for managing correctly this aspect.

9.2.2.7 MSW COLLECTION AREA

The extension of the area that must supply the system thanks to the collection of residual MSW can be easily assessed. It is compulsory that a clear MSW management plan must be previously adopted in order to allow generating reliable scenarios for the present and for the future. Talking about small scale of RDF generation, the area interested to the initiative is expected to be small. That means that the presence of industrial district close to a region with many small towns could couple towns and plants.

9.2.2.8 POPULATION INVOLVED

This parameter is a consequence of the above ones. The initiative does not affect the behaviour of the population of the area. The criteria of collection and the communication to the population will be the same of a conventional approach. The public communication will be simpler than the conventional strategy (centralized direct combustion of residual MSW) as the capacity of the RDF burner serviced to a factory could be very small.

9.2.2.9 CAPITAL COSTS FOR BMT

The capital costs for the construction of a BMT plant do not vary a lot from country to country as the main machines are constructed in few regions and thus the capital costs cannot be strongly affected by a different economical scenario (cheaper manpower). Some differences of capital costs could be anyway associated to a different approach in term of environmental impact minimisation. The process air treatment line can be performed in different ways with different costs (Rada 2005a): the most simple approach is related to the adoption of a bio-filter, while the most expensive solution is based on a thermal regenerative oxidation. In this last case there are also strong differences in term of operating costs because of a consume of methane.

9.2.2.10 OPERATING COSTS FOR BMT

If we consider the cost for manpower in the centre of Europe and we compare it with some areas of the east of Europe we can find differences even of 5 times. That strongly affect the economical balance as the cost for generating one unit of energy as RDF is very different from country to country. Of course the operating costs for bio-drying are only a part of the total costs. Taking into account the reduced capacity of BMT serviced to a specific factory, it could be interesting to integrate the personnel of BMT in the personnel of the factory as the limited amount of waste to be managed and the automation of the management of the process allow adopting a part-time.

9.2.2.11 CAPITAL COSTS FOR RDF BURNER

The capital costs for the construction of a RDF burner do not vary a lot changing country: also in this case the main machines are constructed in few regions and thus the capital costs cannot be strongly affected by a different economical scenario.

9.2.2.12 OPERATING COSTS FOR RDF BURNER

The different scenario of salaries for the workers can affect this item of cost. If the burner is constructed integrated in a factory, some operating costs could decrease thanks to an integrated management of the machines.

9.2.2.13 THERMALLY INTEGRATED SOLUTIONS

Suppose to have an existing factory with an existing natural gas burner. In case of fluctuations in the heat demand, a solution could be based on a steady generation of heat by burning RDF in a new burner and adopting the existing natural gas burner to follow the fluctuations of demand. In this case the integration simplify the management of the plant and guarantee a lower environmental impact as RDF burned by a steady combustion is less impacting. The steady combustion in also guaranteed by the fact that RDF has steady characteristics compared with the original characteristics of waste.

9.2.2.14 WASTE INTEGRATED SOLUTIONS

Until now the scenario has been discussed as based only on the explotation of MSW. In reality, if the area of interest is characterised by a significant generation of special waste with combustible characteristics, the integration of MSW with special waste could guarantee a better economical balance: special waste in some industrial districts could be found with an interesting energy content (wood, packaging, cardboard, etc.).

9.2.2.15 ECONOMIC COMPARISON ANALYSIS

The overall economical analysis can be developed according to the conventional cost-benefits approaches. Some advantages can be taken in countries where special funds (as the European Union structural funds) are available for a co-financing of the plants. Of course, that is not money freely received: some limitations in co-financing are to be taken into account. For instance it is possible to have a support for the BMT plant as it services the population of a region and, may be, the co-financing of a private burning of RDF could not be viable. Additional limitations could be related to the tariff to be applied to the population.

9.2.3 DEFINITION OF A CASE-STUDY

In some regions of the east of Europe there are many industrial districts under construction. The case-study developed in this paper refers to the case of an area with no selective collection (today), with a present generation of 1 kg of MSW per capita, with a LHV equal to 8 MJ kg^{-1}, with many small towns in the surroundings, with an uncorrect MSW management (today). The factory to be considered uses natual gas for generating hot off gas for its industrial process, with a power of 1 MW_{th} and no fluctuation of energy requierement. The manpower is cheap compared to the European averge (300 Euro month^{-1} as a base salary).

TABLE 1: Parameter and results of the analyzed case-study.

parameter	value	units	comments
power$_{needed}$	1	MW_{th}	heat
operation	8000	h y^{-1}	
energy$_{needed}$	8000	MWh_{th} y^{-1}	heat
fuel$_{to\ substitute}$	0.45	Euro Nm^{-3}	natural gas
LHV$_{of\ methane}$	9.3	kWh_{th} Nm^{-3}	
gas$_{required}$	860215.05	Nm^3 y^{-1}	
LHV$_{RMSW}$	8	MJ kg^{-1}	
RMSW$_{production}$	1	kg inh^{-1} d^{-1}	
bio-drying$_{loss}$	28%	-	weight loss
bio-drying$_{energy\ loss}$	3%	-	overall loss
Post-secycling	28%	-	weight loss
LHV$_{RDF}$	15	MJ kg^{-1}	viable as is
ratio$_{RDF/MSW}$	52%	-	
RDF$_{needed}$	1920	t y^{-1}	
MSW$_{needed}$	3692.3	t y^{-1}	
people$_{involved}$	10116	inh.	
methane$_{cost}$	387096.77	Euro y^{-1}	
MSW$_{present\ cost}$	30	Euro t$^{-1}_{MSW}$	Savings
equivalence cost	74.85	Euro t$^{-1}_{MSW}$	RMSW tariff

9.3 RESULTS AND DISCUSSION

In the Table 1 some details and the results of the case-study are presented. It must be pointed out that the approach has been developed "factory ori-ented", in the sense that the final result is the tariff of equilibrium that must be taken into account for veryfing if the cost for treating MSW according to this strategy is competitive with the cost of the present strategy. In the discussed case-study the present strategy is based on the use of natural gas. In practice, the tariff of equilibrium is the one that allow generating a unit of heat from MSW with the same cost of the generation of a unit of heat from natural gas (in different scenarios the conventional fuel could be coal, etc.).

The volume of the bio-container can be easily assessed taking into account the lasting of bio-drying (see Figure 2) and the density of residual MSW: 0.2 t m^{-3}. If only one factory activates this strategy, a second paralle bio-container must be constructed in order to guarantee the acceptance of MSW when the first reactor is full and operating. The cost for thermal energy generation can be lower in case of RDF based strategy if we take into account that the equivalence cost is reasonable: even if, at very small scale, we can increase the influence of personnel costs per unit of waste treated, the tariff seems to be viable. This approach is even more interesting where structural funds can be activated. The substitution of methane gives a higher local impact into the environment for some parameters as the quality of fuel is different. Anyway the local impact can be kept low also thanks to the small stream of waste locally combusted. On the contrary the global impact is lower (a part of RDF is biomass-like).

9.5 CONCLUSIONS

Results show that in areas where the cost for workers is low, the cost for a unit of thermal energy can be lower in case of RDF based strategy and where methane can be substituted. Anyway the method can be adopted to a variety of case-studies. This approach is more interesting for countries where European Union structural funds can be used to support the re-organization of the MSW management. In this case the advantage is higher as these funds can solve the problem of finding financing for the initial investment for covering capital costs, at least for a part of the installations (bio-drying). It is clear that the substitution of methane gives a higher local impact into the environment for some parameters, but the local impact can be anyway kept under acceptable levels. On the contrary the global impact is lower thanks the substitution of methane by RDF. Thus, this approach seems to be more interesting in countries where a target from the Kyoto protocol must be complied with.

REFERENCES

1. Gleis Markus (2009). Technische Lösungen zur Abfallentsorgung in Europa -Strategien und Erfahrungen aus 20 Jahren, Proceedings of Tagung :Gesellschaft, Umwelt und Gesundheit, Ecocenter Bolzano.
2. Grosso M., Rigamonti L., Malpei F (2009). I trattamenti biologici all'interno dei sistemi di gestione integrata dei rifiuti: bilanci energetici ed ambientali, In Compost ed energia da biorifiuti: Vismara, Grossu, Centemero, editore Dario Flaccovio, 63-90
3. Muller Wolfgang (2009). Mechanical Biological Treatment and its role in Europe, Proceedings of Waste–to–Resources 2009, 3rd International Symposium MBT&MRF, Cuvillier Verlag publischer, 1-10.
4. Rada E. C. (2005a). Municipal Solid Waste bio-drying before energy generation, PhD Thesis, University of Trento & Politehnica University of Bucharest.
5. Rada E. C., Franzinelli A., Taiss M., Ragazzi M., Panaitescu V., Apostol T. (2007a). Lower Heating Value dynamics during municipal solid waste bio-drying, Environmental Technology, vol. 28, 463-469.
6. Rada E. C., Istrate I. A., Ragazzi M. (2009b). Trends in the management of the Residual Municipal Solid Waste, Environmental Technology, Vol. 30, No. 7, 651–661.
7. Rada E. C., Ragazzi M., Apostol T. (2008). Role of Refuse Derived Fuel in the Romanian industrial sector after the entrance in EU, Proceesings of Fourth International Conference on Waste Management and the Environment, edited by Zamorano M, 89-96.
8. Rada E. C., Ragazzi M., Apostol T. (2009a). MSW bio-drying in different countries, Proceedings of Waste–to–Resources 2009, 3rd International Symposium MBT&MRF, Cuvillier Verlag publischer, 571-576.
9. Rada E. C., Ragazzi M., Panaitescu V., Apostol T (2006a). MSW bio-drying and bio-stabilization : an experimental comparison, Proceedings of ISWA 2005, International Conference: Towards integrated urban solid waste management system, CD version.
10. Rada E. C., Ragazzi M., Panaitescu V., Apostol T (2005b). Energy from waste: the role of bio-drying, Stintific Buletin, seria C: Electrical Engineering, vol 67, nr. 2, 69-76.
11. Rada E. C., Ragazzi M., Panaitescu V., Apostol T. (2006b). Experimental characterization of Municipal Solid waste bio-drying, Proceedings of Third International Conference on Waste Management and the Environment, Efited by Popov V.,Vol. 92, 295 – 302.
12. Rada E.C., Franzinelli A, Ragazzi M., Panaitescu V., Apostol T (2007b). Modelling of PCDD/F release from MSW bio-drying, Chemosphere, vol. 68, 1669-1674.

13. Ragazzi M, Rada E.C. (2009a). Tecnologie per la bioessiccazione, In Compost ed energia da biorifiuti: Vismara, Grossu, Centemero, Dario Flaccovio editore 307-330.
14. Ragazzi M., Rada E. C. (2009b). MSW bio-drying eco-balance and Kyoto protoco, Proceedings of Waste–to-Resources 2009, 3rd International Symposium MBT&MRF, Cuvillier Verlag publisher, 373-380.
15. Scotti Sergio (2009). Recyclable materials recovery after biological treatment of the residual fraction: quality improuvement and contribution to landfill diversion targets, 3rd International Symposium MBT&MRF, Cuvillier Verlag publischer, 381-388.
16. Shao Li-Ming, Zhang Dong-Qing, He Pin-Jin (2009). Effect of bio-drying on sorting and combustion performances of municipal solid waste, Proceedings of Waste–to-Resources 2009, 3rd International Symposium MBT&MRF, Cuvillier Verlag publischer, 389-399.
17. Tintner J., Smidt E., Meissl K., Binner E (2009). How in MBT-technology in 20 years?, , Proceedings of Waste–to-Resources 2009, 3rd International Symposium MBT&MRF, Cuvillier Verlag publischer, 111-118.

CHAPTER 10

Technical and Economic Efficiency of Utilization of Biogas from Animal Waste for Energy Generation

E. MINCIUC, R. PATRASCU, M. NORISOR, AND D. TUTICA

10.1 THE OPPORTUNITY OF BIOGAS UTILIZATION

Biogas is a combustible gas that is similar to natural gas or to propane-butane gas. The biogas (the biologic gas) traditionally can be obtained after anaerobe fermentation of organic matter.

Due to the global efforts for reduction of emissions of Green House Gases (GHG) and for replacing fossil fuels with renewable energy sources, in the last decade the interest towards the biogas has considerably increased. In addition, there should be mentioned that biogas generation represents an attractive solution for organic waste treatment with great advantages for the environment. In this context biogas and thus energy

Minciuc E, Patrascu R, Norisor M, and Tutica D. "Technical and Economic Efficiency of Utilization of Biogas from Animal Waste for Energy Generation." Printed with permission from the authors.

generation from animal and agricultural wastes represents a priority and needs state support, [1].

The formation of combustible gases due to the process of degradation of organic matter in an environment with molecular oxygen is a process that takes place naturally on our planet. This is the way the underground sediments were created, i.e. natural gas, due to degradation of plants and animals from the oceans and seas of pre-historic times. Identically, on the Earth surface there are formed combustible gases anywhere where there is organic matter that are in anaerobic conditions, [2].

The calorific value of biogas depends on non-combustible elements (N, CO_2); their presence reduces the energy quality of biogas. Lower Calorific Value of biogas varies between 20 and 24 MJ/m^3, [3].

TABLE 1: General characteristics of biogas, [4]

Composition	55-70 % methane
	30-45 % carbon dioxide
	Other gases
Energy content	6.0 – 6.5 kWh/ m^3
Fuel equivalent	0.6- 0.65 l oil/m^3 biogas
Explosion limits	6-12 % biogas in the air
Ignition temperature	650 – 750 °C
Critical pressure	75-89 bar
Critical temperature	82.5 °C
Density	1.2 kg/ m^3
Odor	Rotten eggs (treated biogas has almost no odor)
Molar mass	16,.043 kg/mole

10.2 THE DOMAIN OF VARIATION OF CALORIFIC VALUE

Generally speaking, the biogas is a mixture of gases—methane, carbon dioxide, nitrogen, carbon monoxide, etc. The higher the methane weight in the biogas the higher is the energy quality of the fuel. Carbon dioxide is a neutral gas, it is not combustible, but the carbon monoxide is a combustible gas. The composition of generated biogas depends very much on the type of biomass used as raw material and on the technology used for biogas production, [1].

Typical composition of biogas, [5, 6, 7]:

- 40-75 % Methane, CH_4;
- 25-60 % Carbon dioxide, CO_2;
- 0-10 % Nitrogen, N_2;
- 0-1 % Hydrogen, H_2;
- 0-3 % Hydrogen sulphide, H_2S.

TABLE 2: Energy value of biogas compared to other fuels

Fuel	Unit	Calorific value kcal/unit	Equivalent of unit for 1 m^3 of biogas
Biogas with cu 60 % methane, 0 °C, 1 bar	m^3	5130	1
Fresh wood	kg	1300-1800	3.95-2.85
Dried wood	kg	1800-2200	2.85-2.34
Lignite	kg	1800-3800	2.85-1.35
Coal dust briquettes	kg	4000-6800	1.28-0.76
Oil fuel	kg	9400-9500	0.55-0.54
Diesel	kg	10000-11000	0.51-0.47
Natural gas methane	m^3	8500	0.60
Liquefied petroleum gas	m^3	22000	0.23

10.3 SOURCES FOR BIOGAS GENERATION

Biogas can be generated from any organic matter of vegetal or animal origin, including wastes. The raw material used for biogas generation can be of different origin, [1]:

Biomass from agriculture:

- Different herbage, lucerne.
- Wastes (tree leafs, roots, straws, etc.).
- Plants (sunflower, sugar beet, etc.) and their silage.

Livestock products:

- Animal manure.

Municipal organic wastes:

- Water and sewage sludge.
- Municipal solid organic wastes.

Organic wastes of industrial origin:

- Food wastes (from restaurants, canteen, etc.).
- Wastes and residues from agro-food industry.
- Wastes from production processes of food industry.
- Fats.

The biogas that is generated using the process of anaerobic digestion is cheap and is a renewable energy source. In the combustion process of biogas a neutral CO_2 is generated. The anaerobic digestion also offers the possibility of treatment and recycling of different types of wastes, such as agricultural wastes, sewage waters and sludge, industrial waste water, etc. using a sustainable and environmentally friendly solution, [8].

10.4 CASE STUDY: ANALYSIS OF EFFICIENCY OF UTILIZATION OF BIOGAS FROM A PIG FARM IN A COGENERATION PLANT

10.4.1 DESCRIPTION OF THE ANALYZED CASE STUDY

Technical and economic analysis has as starting point the fact that animal waste treatment is compulsory and, thus, it is desired to find an advantageous solution for energy valorization of generated biogas.

The reference solution, for technical and economic evaluation of analyzed biogas generation solution it is considered that the agricultural wastes are treated in specialized tanks and generated biogas is flame burned. For covering energy demands of the farm electricity is bought from the national power grid, and heat is generated locally in a natural gas fuelled boiler.

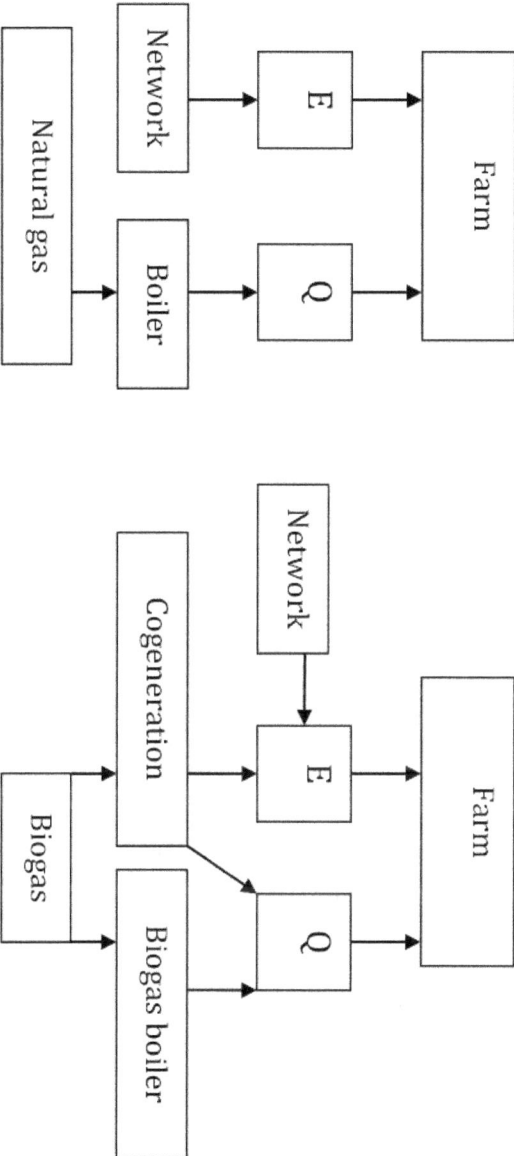

FIGURE 1: Reference solution (a) and biogas cogeneration solution (b) Notations used: E – electricity (consumed at the farm) and Q – heat (consumed at the farm)

The analyzed cogeneration solution involves installation at the farm of a anaerobic digester, a storage tank, a biogas treatment plant and an internal combustion engine that operates on biogas. When the internal combustion engine is not operated the biogas is combusted in a boiler and the surplus is flame burned.

10.4.2 DETERMINATION OF DAILY BIOGAS ENERGY POTENTIAL

The waste quantity that comes from an animal species depends on many factors, such as: animal category, applied keeping system (indoor, outdoor, etc.), existence or non-existence of bedding, type of used bedding, floor type (with or without grate), the way of watering the animals and cleaning the floor, type of forage used to feed animals, which leads to different water consumption, the way of feeding the animals, [1].

The daily biogas flow is:

$$D = SV \cdot ef \cdot d_{sv} \cdot N \cdot \varepsilon \ [Nm^3/zi] \tag{1}$$

where:

- N: number of animals
- SV: quantity of volatile solids [kg/day animal]
- ef: efficiency
- ε: efficiency of wastes collection
- d_{sv}: flow of destroyed SV [Nm³/kgSV]

The calorific value of biogas is:

$$PCI_b = c_i^{CH4} \cdot p_{CH4} \ [Wh/Nm^3] \tag{2}$$

where:

- p_{CH4}: specific calorific value $[Wh/\%CH_4]$
- c_i^{CH4}: methane concentration [%]

Energy generated daily through anaerobic digestion of animal wastes is:

$$B = D \cdot PCI_b \; [Wh/zi] \qquad (3)$$

10.4.3 DETERMINATION OF DAILY ENERGY CONSUMPTIONS

The estimated values of maximum specific energy consumptions are:

- Electricity specific consumption e= 6.95 [We/cap]
- Heat specific consumption q= 16.65 [Wt/cap]

Daily electricity quantity consumed during winter days:

$$E = e \cdot N \; [kWe/day] \qquad (4)$$

Electricity consumption for spring-autumn days has been considered as being equal to 0.7 from the winter days consumption, and for summer days electricity consumption has been considered as being equal to 0.4 from winter days consumption.

Daily heat quantity

$$Q = N \cdot q \; [kWT/day] \qquad (5)$$

The heat consumption during spring-autumn days has been considered as being equal to 0.5 from the consumption of winter days, and for summer days has been considered equal to 0.2 from the consumption of winter days.

The number of days taken into consideration for the calculus is as follows:
- Winter days : 150 days/year
- Spring-autumn days: 125 days/year
- Summer days: 90 days/year

The quantities of electricity and heat that are consumed annually at the farm have been determined through multiplication of daily consumption with typical number of days.

There are taken the following hypotheses:

- The own electricity consumption of the farm has been estimated as being 60 % from the total consumption; the rest is needed for biogas generation and utilization.
- The own heat consumption of the farm has been estimated as being 50 % from the total consumption; the rest is needed for biogas generation and utilization.

10.4.4 EQUIPMENT SIZING FOR BIOGAS VALORIZATION

The volume of the digester needed for anaerobic fermentation of wastes is:

$$V = \lambda \cdot v \cdot N \cdot \varepsilon \ [m^3] \tag{6}$$

where:

- λ: coefficient of over-sizing the digester
- v: specific volume [m^3/animal]

The size of the internal combustion engine and boiler that use biogas is determined taking into account the daily energy content of produced biogas.

The biogas utilization module has the following priority phases:

1. Combustion in internal combustion engine
2. Combustion in boiler

3.   Storage
4.   Flame burning

The installed electric power of the internal combustion engine is:

$$P_{coge} = B \cdot \eta_e / 24 \text{ [We]} \tag{7}$$

where:

-   η_e: electric efficiency of the cogeneration unit

During the periods when the engine is not operated biogas shall be used in boiler. The maximum capacity of the boiler is:

$$q_{caz} = B \cdot \eta_{caz}/24 \text{ [Wt]} \tag{8}$$

where:

-   η_{caz}: boiler's efficiency

A basic condition is to use biogas without reselling the excess of electricity (case A).
The biogas consumption in the cogeneration unit is:

$$b_{mot} = P_{el} / \eta_{el} \text{ [W]} \tag{9}$$

The biogas consumption in the boiler is:

$$b_{caz} = q_{caz} / \eta_{caz} \text{ [W]} \tag{10}$$

The annual electricity generation in cogeneration unit is:

$$E_{an} = P_{el} \cdot T_{an,e} \quad [kWh/year] \tag{11}$$

where:

- $T_{an,e}$: number of operating hours during the year at maximum load

The annual heat generation in the boiler is:

$$Q_{caz,an} = Q_{caz} \cdot T_{an,caz} \quad [kWh/year] \tag{12}$$

10.4.5 TECHNICAL AND ECONOMIC ANALYSIS

From economic point of view there have been analyzed 2 different solutions:

1. An integrated plant: For this solution it has been considered that wastes are treated and produced biogas is used for covering a part farms own needs. Produced electricity is consumed on-site. The electricity peak demand is covered from the national power grid. In this case the company shall not obtain any green certificates.
2. Separate plant: For this solution it has been considered installation of a CHP plant and a biogas boiler, which will produce heat for farm and digesters, and whole amount of electricity shall be sold into the national power grid, thus obtaining green certificates.

The analysis of economic efficiency has been performed based on technical and economic criteria using discounted values:

1. Net Present Value, NPV: is the most comprehensive and observable indicator of economic efficiency. It includes all economic efforts and effects that have been realized during the project lifetime.
2. Internal Rate of Return, IRR: it reflects the project's capacity to generate profit, offering clues about acceptability of loans. IRR represents

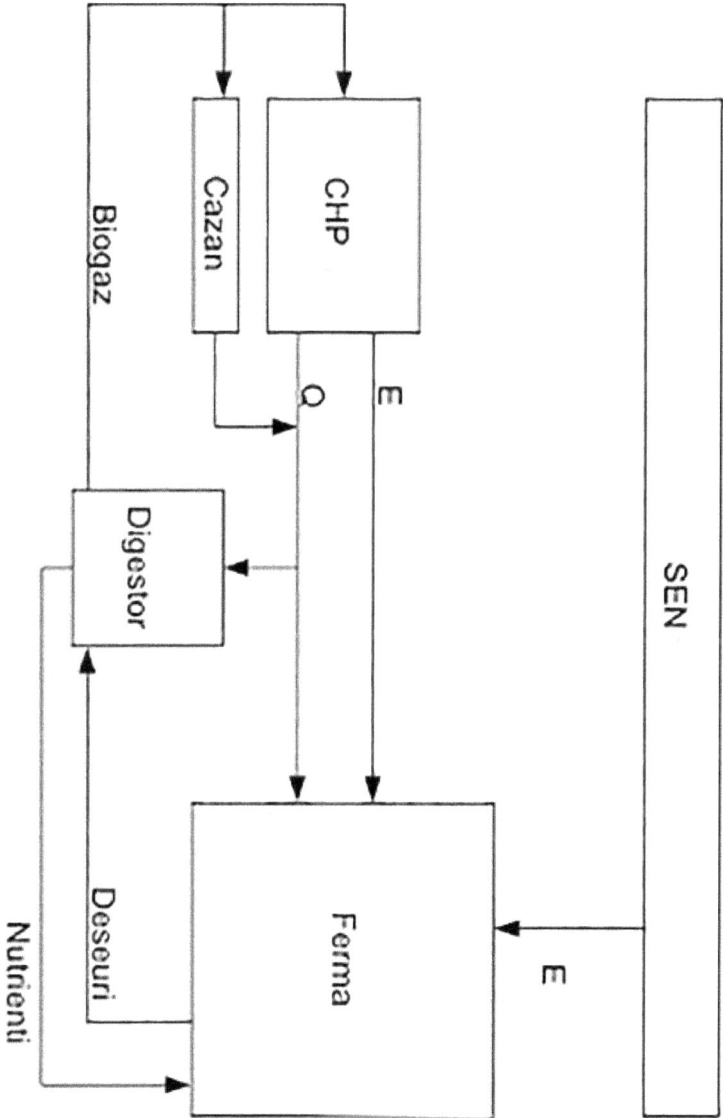

FIGURE 2: General layout of the integrated plant

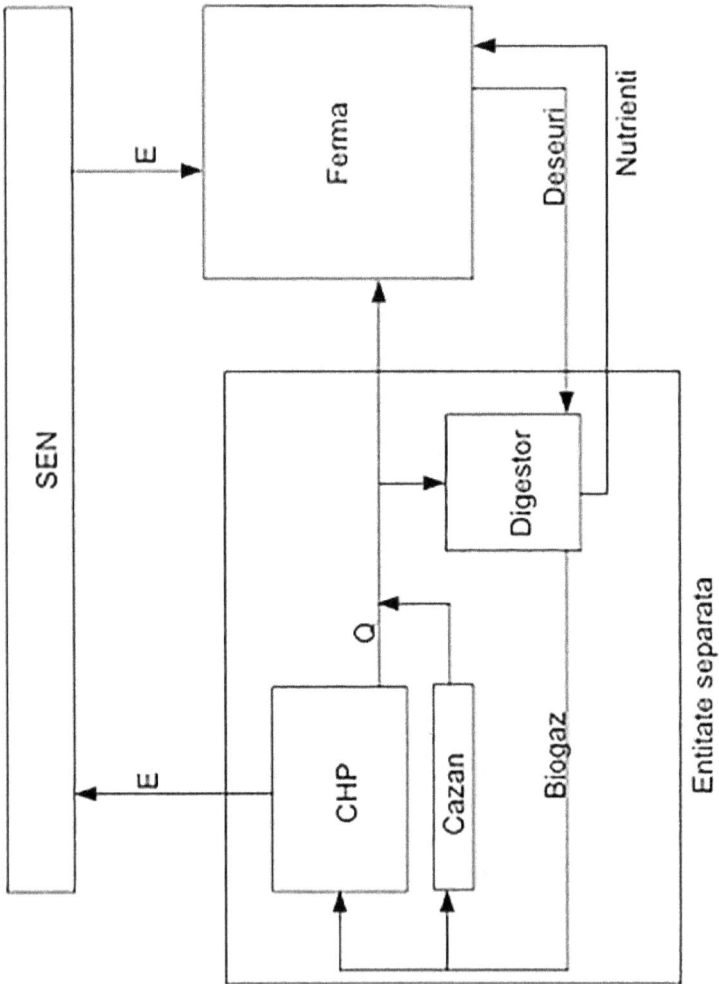

FIGURE 3: General layout of separate plant

the value of the discount rate for which NPV is zero. For the case IRR is greater than discount rate the project is considered to be feasible.
3. The Discount Payback Period, DPP: it represents the period of investment return using discount values.

For determination of the above mentioned criteria there have been considered the following elements:

Investment Estimation
There have been considered investments in the following equipment: digester, CHP unit, biogas boiler, and biogas storage tank and filtration equipment.

The investment in the digester is calculated by multiplying the specific investment for digester, i_{dig} with digester volume:

$$I_{dig} = i_{dig} \, V \; [\text{€}]$$
(13)

The investment in the cogeneration unit is estimated using specific investment and installed capacity:

$$I_{cog} = i_{cog} \, P_{cog} \; [\text{€}]$$
(14)

The investment in the biogas boiler is estimated using specific investment:

$$I_{caz} = i_{caz} \, q_{caz} \; [\text{€}]$$
(15)

For the filtration equipment the investment depends on the maximal biogas flow that is consumed by internal combustion engine:

$$I_{filt} = i_{filt} \, b_{mot} \; [\text{€}] \tag{16}$$

The investment in the biogas storage tank is calculated using specific investment and the tank volume.

$$I_{stoc} = i_{stoc} \, V_{stoc} \; [\text{€}] \tag{17}$$

The total investment for the biogas plant shall be the sum of all investments in the different equipment needed for the process:

$$I_{total} = I_{dig} + I_{cog} + I_{caz} + I_{filt} + I_{stoc} \; [\text{€}] \tag{18}$$

Estimation of the Maintenance Costs

These costs are estimated taking into consideration specific maintenance costs for different equipment.

Estimation of Expenses with Fuel and Electricity

For the cogeneration solution the expenses with fuel include expenses for natural gas purchase needed for start up the cogeneration unit that then generates heat for heating up the digester till biogas is generated; these expenses are negligible during the summer time.

For the case (A) there is a need for a certain amount of electricity bought from the national power grid for covering up all farm's needs. For increasing economic efficiency of the cogeneration unit there should be used first all electricity generated by it and then to buy the difference from the grid.

The annual sum allocated for these expenses shall the electricity quantity purchased from the national power grid multiplied by electricity tariff.

For the case (B) it is necessary to buy all the amount of electricity from the national power grid for covering farm's needs and for biogas generation; it is due to the fact that all electricity generated by CHP unit is sold into the national power grid with the purpose of getting green certificates.

Estimation of Revenues

1. For the case of the integrated plant the annual revenues of the project represent the savings due to power and heat generation on-site using biogas.
Annual savings due to electricity generation on-site are:

$$VA_{el} = t_{el}\ E_{an} \tag{22}$$

Annual savings due to heat generation on-site using biogas are:

$$VA_t = t_t\ (E_{t,cog} + E_{t,caz}) \tag{23}$$

The total annual revenues/savings shall be the sum of the previous two:

$$VA_{tot} = VA_{el} + VA_t \tag{24}$$

2. For the case of separate plant the annual revenues represent the incomes from electricity selling into the grid and from selling green certificates.

10.4.6 HYPOTHESES USED FOR THE ANALYSIS

For the technical and economic analysis there have been used the following hypotheses:

- There has been considered a farm with 10000 animals.
- The daily biogas flow at the farm is considered to be constant.
- The discount rate is 10%.
- The number of days used for annual energy consumption calculation have been considered as follows:

- Winter days: 150
- Spring-autumn days: 125
- Summer days: 90
- The capacity of the storage tank shall ensure the stocking of biogas generation for 6 hours. The biogas storage equipment shall allow a uniform biogas flow and high biogas quality.
- The installed capacity of the boiler shall be chosen so the boiler can burn˙as much biogas as possible during the periods when the internal combustion engine is not operated, but without producing more heat than necessary for the farm.
- Biogas valorization generated through waste treatment involves biogas treatment. There shall be installed biogas treatment equipment that are sized depending in the flow of treated biogas.

10.5 RESULTS OF THE ANALYSIS FOR A PIG FARM

The results of the technical and economic analysis for a 10000 pig farm are shown in table 3.

TABLE 3: Results of technical and economic analysis for a pig farm

Item		Unit	Value case A	Value case B
Number of animals		-		10000
Efficiency of waste collection		%		95
Specific daily flow of biogas		Nm³/day animal		0.12
Average methane concentration		%		58
Lower calorific value of biogas		kWh/m³		6.09
Daily biogas flow		Nm³/day		1117.2
Daily energy potential of biogas		kWh/day		6803.75
Hourly biogas energy		kW		283
Digester volume		m³		1596
Engine electric installed capacity		kWe		112.5
Engine heat installed capacity		kWt		143.8
Boiler installed capacity		kWt	250	
Engine operation time	winter	h/year	2225	4170
	spring-autumn	h/year	1300	1750
	summer	h/year	500	500

TABLE 3: *Cont.*

Item		Unit	Value case A	Value case B
Boiler operation time	winter	h/year	1120	0
	spring-autumn	h/year	250	0
	summer	h/year	0	0
Annual quantity of generated electricity		MWhe	452.812	722.250
Annual quantity of heat generated in engine		MWht	578.795	923.196
Annual quantity of heat generated in boiler		MWht	342.500	0
Biogas surplus flame burned annually		m³		
Annual amount of electricity bought from grid		MWhe	3.83	0
Average tariff of bought electricity		€/MWhe		70
Heat tariff in network		€/MWht	40	-
Tariff for selling heat to farm (B)		€/MWht	-	30
Tariff for selling electricity to grid		€/MWhe	-	40
Minimum price of green certificate		€/CV	-	28.8
Digester investment		Thousands €		164.39
CHP unit investment		Thousands €		90.90
Boiler investment		Thousands €		23.75
Filtration equipment investment		Thousands €		6.14
Storage tank investment		Thousands €		33.00
Works		Thousands €		40.00
Cost for capital repair discounted (B, year 7)		Thousands €	-	12
Total investment		Thousands €	358.18	370.18
Annual maintenance costs (engine+boiler)		Thousands €	4.65	5.78
Annual costs with electricity purchase		Thousands €	0.267	-
Annual cost with electricity for own needs (B)		Thousands €	-	12.68
Total annual expenses		Thousands €	4.92	18.46
Annual savings due to electricity		Thousands €	19.02	-
Annual savings due to heat		Thousands €	18.43	-
Annual revenues due to green certificates		Thousands €	-	62.48
Annual revenues due to selling of electricity		Thousands €	-	28.89
Annual revenues due to selling heat to farm			-	13.85
Total annual revenues		Thousands €	37.44	105.14
Net annual revenue		Thousands €	32.52	86.68
Discount rate		%		10
Project lifetime		years		10

TABLE 3: *Cont.*

Item	Unit	Value case A	Value case B
Total discounted costs	Thousands €	388.43	483.7
Total discounted revenues	Thousands €	230.28	646.61
NPV	Thousands €	<0	162.91
IRR	%	-	19 %
DPP	years	-	2.3
Grant need for IRR 12 %	%	50	-

10.6 CONCLUSIONS

The biomass and waste utilization can be justified especially due to ecological considerations, due to the need of waste treatment.

For increasing biogas production there can be combined different types of wastes/raw materials. There can be mixed different animal manure and also different vegetal substances.

Within the context of sustainable development, where the energy and environmental issues should be solved simultaneously, biogas production and its utilization as an energy source is of great interest due to many advantages, such as, [9]:

1. Energy generation.
2. Waste treatment.
3. Stabilization of animal manure.
4. Reduction of pathogen loads.

The feasibility of implementation of biogas equipment, generally, depends on many key factors:

• Biogas production scale.
• Availability of energy crops and their specific costs.
• Waste production at analyzed farm.
• Policies and regulations regarding waste treatment.
• The value of bio-fertilizers.
• Possibility of local use of heat.

- Reasonable costs for investments.
- Availability of equipment and buildings that can be integrated in the biogas plant concept.
- Availability of grants.

The results of the analyzed cases lead to the following conclusions:

- The heat demand of agricultural farms can be integrally covered by biogas generated through waste treatment.
- With no financial facilities a CHP biogas project for agricultural farms can hardly be feasible from the economic point of view.
- In situation with green certificates case B is the best one.
- For the case when green certificates award for the next 10 years is not certain it is preferable the solution with financial grants.

The performed analysis established a set of criteria that should be analyzed when implementing such a solution:

- The biogas unit should be placed near to raw materials to minimize the costs.
- The biogas unit should be placed near to town so the heat can be sold.
- More than 70 % of raw material should come from own sources or there should contracts for supply for at least 10 years.
- The raw material flow should quiet stable.
- The utilization of secondary product, digestate, can significantly increase economic feasibility of the project.

REFERENCES

1. Valentin Arion, Olga Şveţ şi Constantin Borosan, Producerea biogazului din deşeuri animaliere, Universitatea Tehnică a Moldovei, Chişinău 2013
2. Tudor Ambros, Surse regenerabile de energie , editura „Tehnica –info",1999
3. Adrian Badea, Horia Necula, Surse regeerabile de energie, Agir Bucuresti 2013
4. Ing.Adrian Eugen Cioabă ,Contribuţii teoretice şi experimentale privind poducerea de biogaz din deşeuri de biomasă, teză de doctorat 2009
5. McKendry Peter, Energy production from biomass:Conversion technology, Bioresource Technology, 2001
6. http://www.nikolicivasilie.ro/lucrari-stiintifice/Biogaz%20curs.pdf ;
7. Ioan Bitir-Istrate,Eduard Minciuc, Valorificarea biogazului pentru producera energiei electrice si termice, Cartea Univrsitară,2003
8. http://www.gazetadeagricultura.info/eco-bio/565-energie-regenerabila/14414-tehnologii-actuale-de-obtinere-a-biogazului.html
9. http://www.probiopol.de/10-Care-sunt-componentele-une.45.0.html?&L=1

10. Augustin Ofițeru, Mihai Adamescu, Florian Bodescu,Biogazul ghid practic, 2008

11. Ioan Paunescu,Gigel Paraschiv, Managementul mediului si obtinerea biogazului în fermele suinicole, Facultatea de Ingineria Sistemelor Biotehnice - UPB

12. Valentin Arion, Olga Șveț și Constantin Borosan,Utilizarea biogazului la producerea căldurii și electricității, Chișinău, 2013

13. V. Arion, dr.hab., prof.univ., C. Gherman, T. Tutunaru ,Fezabilitatea economico-financiară a producerii energiei electrice și termice la mini-cet prin valorificarea biogazului , Universitatea Tehnicăa Moldovei,2009

14. Dominik Rutz,Utilizarea durabilă a energiei termice a instalațiilor de biogas, 2012 WIP Renewable Energies, Munchen, Germania, 2012

15. Adrian Eugen Cioabla, Ioana Ionel, Biogazul, energie pentru viitor, editura Politehnica , 2011

16. Victor Emil Lucian ,Surse alternative de energie, Bucuresti, matrix Rom 2011

17. Ing. Nicolae Lontiș, Cercetări teoretice și experimentale privind cogenerarea cu motot M.A.I., funcționînd cu biocombustibil, teză de doctorat, 2008

18. Robert Pecsi,Utilizarea biogazului ca sursă de energie electrică, Facultatea de Instalatii, UTCB

19. Dr Ralf Utermohlen ,Măsuri, metode și soluții pentru poligenerarea industrială a Biogazului în România. (AGIMUS) Germania , martie 2009

20. Screening the rise of fermentable wastes & market prices for energy and waste treatment in Romania, Research report carried out by SC Project Developer (ProDev), Romania, December 2008

21. Michael Kuttner, dr. Sigrid Kusch, Achim Kaiser, dominik Durrie, Economic modelling of anaerobic digestion/ biogas installations in a range of rural scenarios in cornwall, prepared byinternational biogas and bioenergycentre of competence, 21 august 2008

CHAPTER 11

Potential of Bio-Drying Applied to Exhausted Grape Marc

ELENA CRISTINA RADA AND MARCO RAGAZZI

11.1 INTRODUCTION

The wine industry produces every years in the world about 270 millions of hectolitres of wine, of which about the half in European Union (EU), and in particular in Spain, Italy and France; Italy and France compete each year for being the main world wine producer.

The central wine regulations in the EU is entitled *Council Regulation on the common organization of the market in wine,* of 29 April 2008 and was supplemented by *Council Regulation (EC) No 491/2009* of 25 May 2009 amending Regulation (EC) No 1234/2007 that establish a common organization of agricultural markets and specific provisions for certain agricultural products. From the regulatory point of view the wastewater generated from the sector must be treated according to the principals available for the management of this kind of discharge.

This article is an original article. Reprinted with the authors' permission.

The Province of Trento, Italy, where the present research was performed, generates about 800,000 hectolitres of wine per year. To produce one liter of wine about 1.3 kg of grape are utilized, and about the 20% of this grape represents a 'waste' of the wine industry (1). The exhausted grape marc (GM) or spent GM, is the material that results after fermentation and distillation of the grapes in the wine companies and has a high organic content (2). This exhausted grape marc contains generally about 72% skins, 20-22% seeds and 7-8% stalks (3). The exhausted GM used for the experimentation is presented in Figure 1.

Grape marc has traditionally been used to produce pomace brandy (such as grappa), grape-seed oil, polymers and seldom as supplement in animal feed (4,5,6). Today, it is mostly used as fertilizer or for producing renewable energy. However, in 2011, Devesa-Rey presented a study regarding the possibility of valorization of winery waste vs. the costs of not recycling (2). The valorization scenarios didn't take into account the potential role of GM bio-drying. This lack confirms the originality of the contents of the present paper, where an alternative technique to treat exhausted grape marc is presented for energy purposes. The chosen process is an aerobic biological-mechanical treatment: the bio-drying one. The aim of this process is the exploitation of the exothermic reactions for the evaporation of the highest part of the water present in the treated material with the lowest conversion of organic carbon.

The experimental results of a pilot scale bio-drying process applied to exhausted grape marc are presented and discussed in this paper. An important advantage of the use of bio-drying is that the final product, like Solid Recovered Fuel (SRF), is considered biomass opening to the incentives market for renewable energy.

An important parameter for the SRF characterization is the potential rate of microbial self heating. This rate can be determined by the real dynamic respiration index (RDRI = average value of the respiration indexes representing 24 h showing the highest aerobic microactivity). Until 2010 in Italy this parameter has been requested only for the stabilized organic fraction and not for fuels (7). With the transposition of the technical norm CEN/TR 15590 in the decree 205/2010, this RDRI parameter became a key parameter for SRF (8).

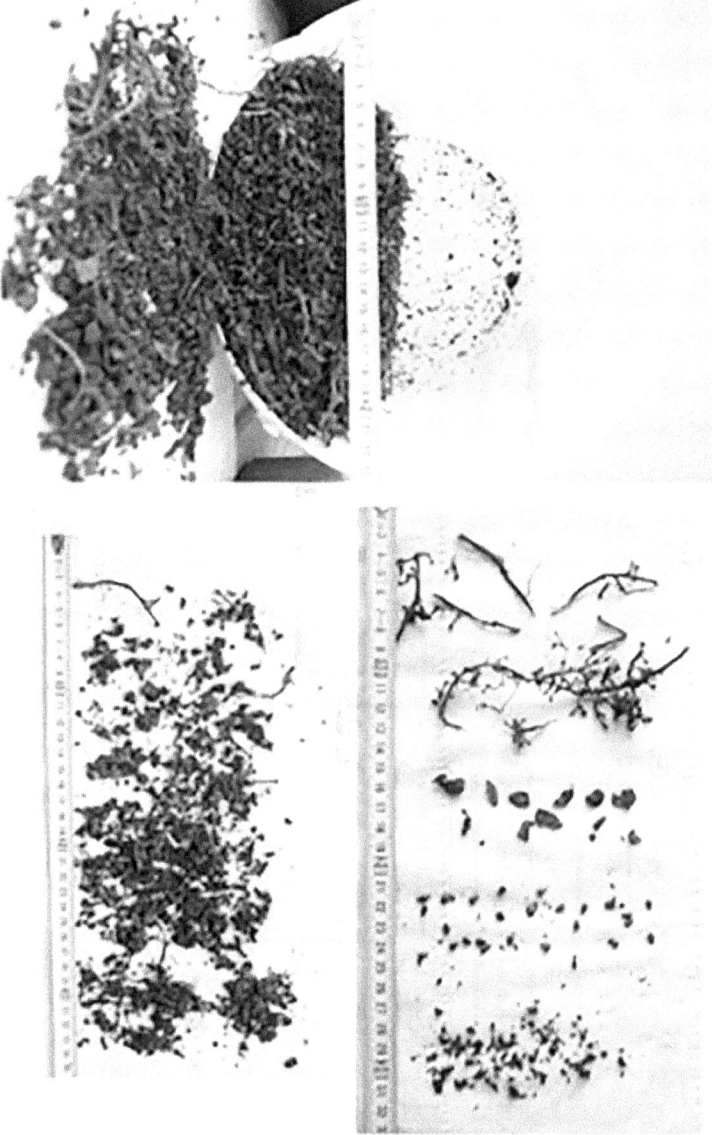

FIGURE 1: Exhausted grape marc: seeds, skins and stalks

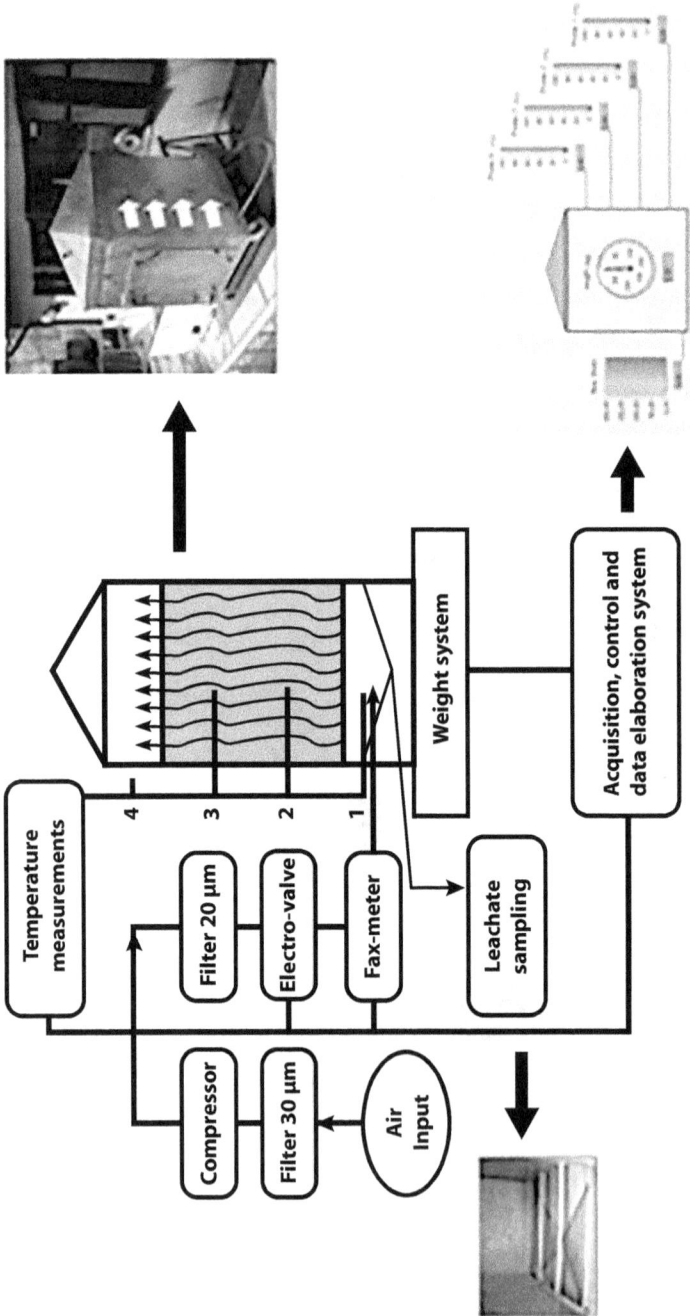

FIGURE 2: Biological reactor

Bio-drying of exhausted grape marc can be an alternative to thermal drying generally performed in the sector by an integrated plant of thermal-pre-treatment and combustion. This integration has the disadvantage of requiring a centralization of the pre-treatment. Grape marc bio-drying could open to a decentralized pre-treatment of this organic substrate before a centralized combustion.

11.2 DESIGN AND METHODS

In order to develop the bio-drying process, a bioreactor of $1\,m^3$ was used at University of Trento (9,10). When performing a run, the adopted reactor (Figure 2) was placed on an electronic balance for monitoring the mass loss during the bio-drying process. The process air was sent into the reactor through a steel diffuser, placed at the bottom. The air crossed upward the matter from the lower part, activating the biological reactions and went out of the biological reactor from the upper part, extracting a part of the water content of the material. The leachate was discharged continuously in order to avoid alterations of the dynamics of weight loss. For monitoring the temperature during the bio-drying process, four temperature probes were placed inside the reactor, one at the air outlet/inlet and two at intermediate levels. All these equipments were connected to a data acquisition system developed for a good management of the process.

A bio-chemical model based on an energy and mass balance was used for a correct interpretation of the experimental data reported in this paper (9). In particular the model reconstructs the composition of the bio-dried material through a process balance taking into account the mass, the bio-chemical reactions stoichiometry and the available energy, resulting at the end a determined mathematical system that has as input the quantity of water, carbon, hydrogen, oxygen and nitrogen contained in the initial mass of exhausted GM and as output the amount of water, carbon, hydrogen, oxygen and nitrogen consumed/removed during the bio-drying process. The model also describes the dynamics of the calorific value during the bio-drying process. Additionally the model gives the dynamics of the following parameters of interest:

- internal energy consumption (overall available Lower Heating Value, LHV);
- volatile solids reduction (assessed from an energy balance);
- net moisture extraction (taking into account the role of hydrogen);
- highest theoretical NH_3 emission factor (supposing that 100% of nitrogen is converted into ammonia).

The aeration criteria were adopted in order to optimize the temperature of the biological bio-drying process taking into account the microbial activity needs: the highest value was limited at 65°C.

For the experimental run, that lasted 30 days, about 96 kg of exhausted grape marc were used. The height of the material in the bio-drying pilot plant was about 22 cm. The moisture content of the exhausted grape marc and of the bio-dried one was determined using a stove at 105°C for 24 hours.

For the volatile and total solids (VS and TS) content and also for the elemental composition a provincial laboratory was used. Carbon, Hydrogen, Oxygen, Sulfur and Nitrogen were determined taking into account the UNI 9903-6:1992 norm, and for the Copper content the EPA 3050B and EPA 6010C were used. For the Chlorine and Mercury content needed for discussing the role of bio-drying towards SRF generation, the determination was made taking into account the ISTISAN'99 and EPA 50505 or/and EPA 9056A.

Generally the lasting of the bio-drying runs is about 2 weeks, when adopted for municipal solid waste (11). For a better understanding of the potential of the bio-drying process applied to exhausted grape marc:

- lasting of was doubled;
- data from respirometry tests regarding the oxygen consumption of the exhausted grape marc before and after the bio-drying process, were generated;
- Chemical Oxygen Demand, COD, was determined in specific cases.

The dynamic respirometer AIR NL (Respirometric Index Analyzer with Not Limiting oxygen) developed by the University of Trento was used in order to have data about the Oxygen Uptake Rate of the organic matrix (12). Inside of this instrument a controlled aeration in no limiting oxygen conditions is made. The Respirometric Index (IR= potential rate of microbial self heating) calculation starts from the decrease of the oxygen concentration curve in the interstitial air of the waste sample, due to bacterial activity. This rate can be determined by the real dynamic respiration index (IR_{24} = RDRI = average value of the respiration indexes representing 24 h showing the highest aerobic microactivity).

11.3 RESULTS

In Table 1 data regarding VS and TS of the exhausted grape marc before bio-drying are presented. Table 2, instead, presents data regarding the ratios VS/TS and TS/GM, the moisture of the initial exhausted GM and the moisture of the bio-dried GM. These values are similar with the ones available in the literature (13). The chlorine and mercury contents were measured: the values for the used exhausted GM sample resulted Cl < 50 mg/kg and Hg = 0.0022 mg/kg respectively.

TABLE 1: Elemental composition of initial exhausted grape marc [%TS and % VS]

	Carbon	Hydrogen	Oxygen	Nitrogen	Sulfur	Ash
% of TS	54.94	5.38	32.08	2.09	0.21	4.85
% of VS	57.74	6.13	33.72	2.20	0.22	-

TABLE 2: VS/TS, TS/GM ratios and moisture and copper content

	kg_{VS}/kg_{TS}	kg_{TS}/kg_{GM}	kg_{H2O}/kg_{GM}	$mg_{Cu}/kg_{bio-driedGM}$
Initial exhausted GM	93.06%	34.94%	65.06%	-
Bio-dried GM	81.38%	47.08%	52.95%	130

Using the cited bio-chemical model the dynamics of the main parameters for characterizing the exhausted grape marc bio-drying were developed (9). Data concerning temperature dynamic during the bio-drying run are reported in Figure 3.

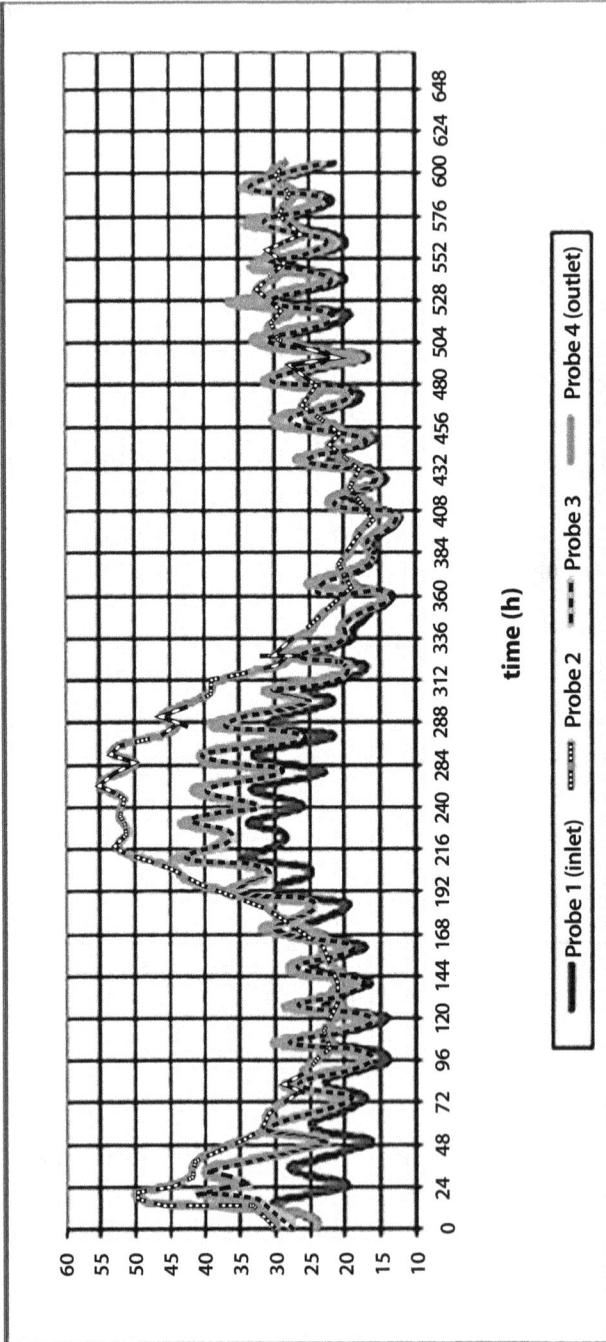

FIGURE 3: Temperature dynamics during the bio-drying run

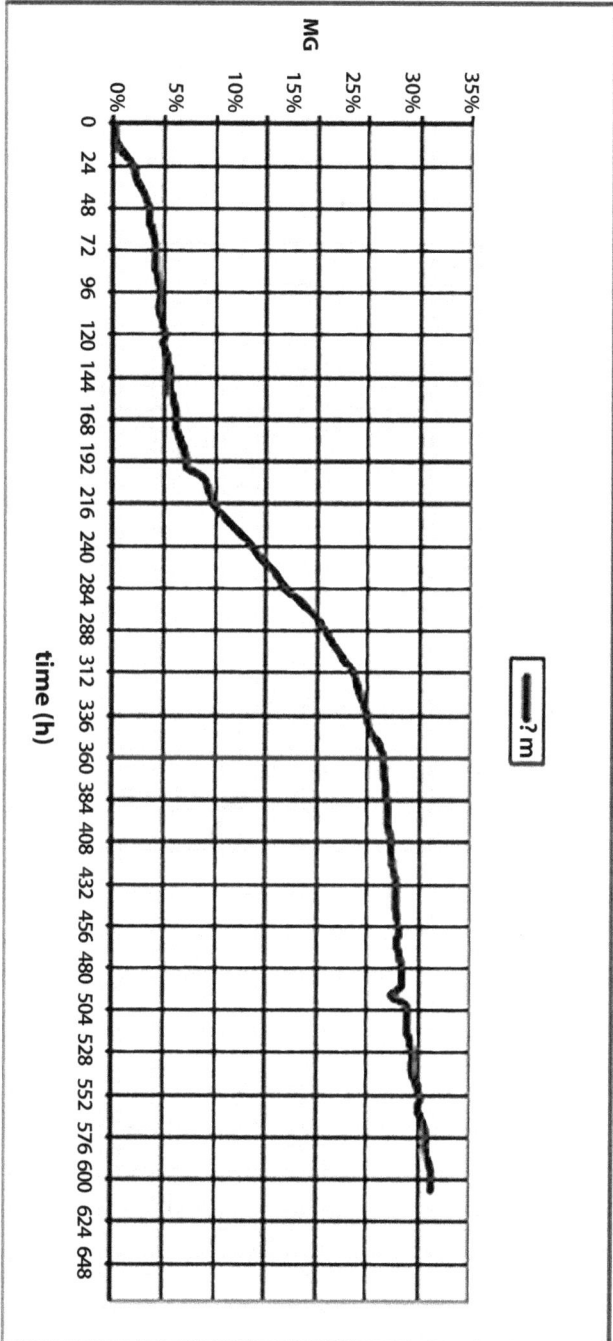

FIGURE 4: Mass loss dynamics during the bio-drying run

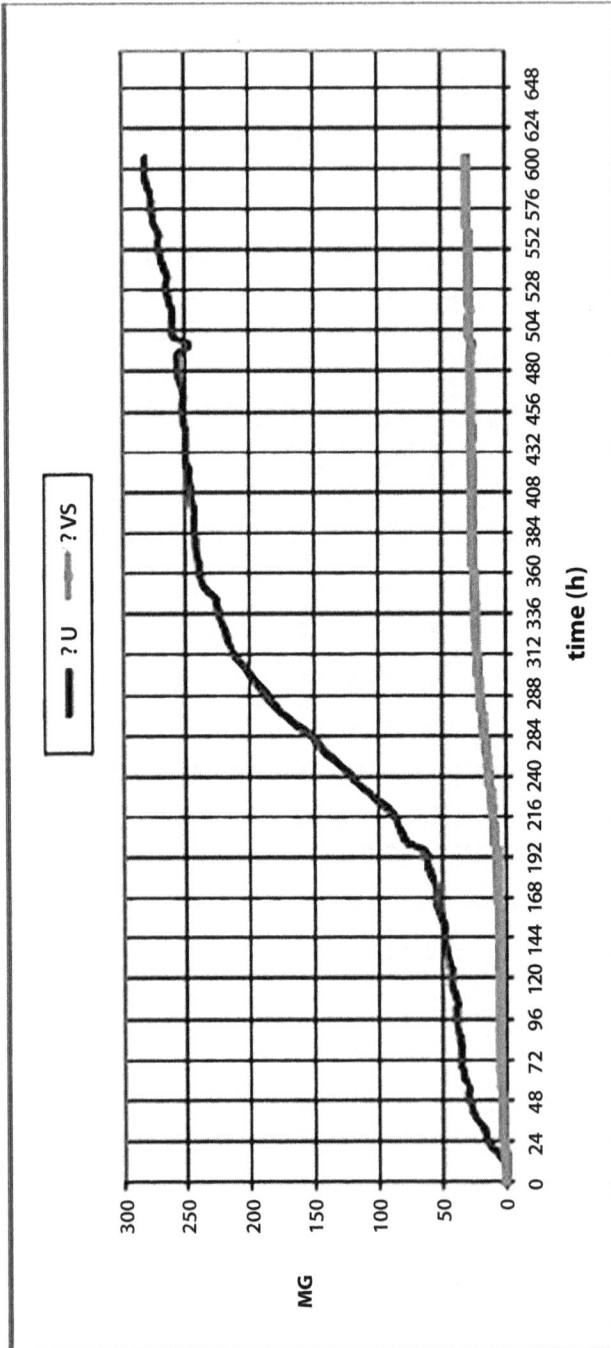

FIGURE 5: Moisture and volatile solids dynamics (ΔU and ΔVS) during the bio-drying run

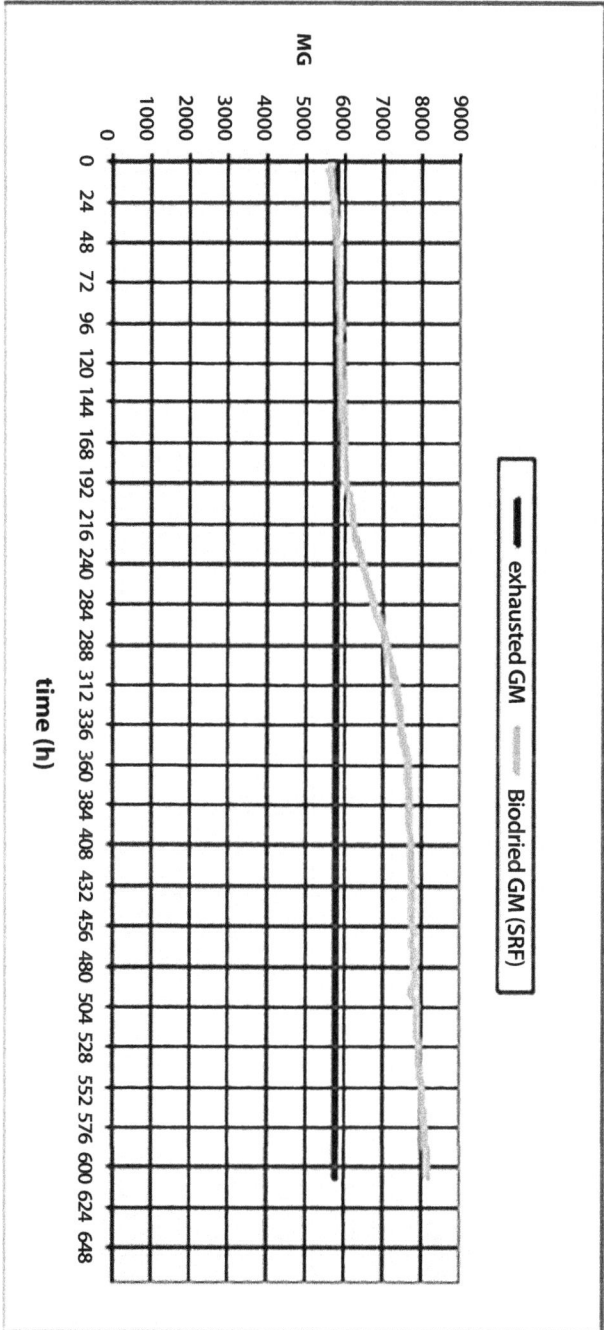

FIGURE 6: LHV dynamics during the bio-drying run

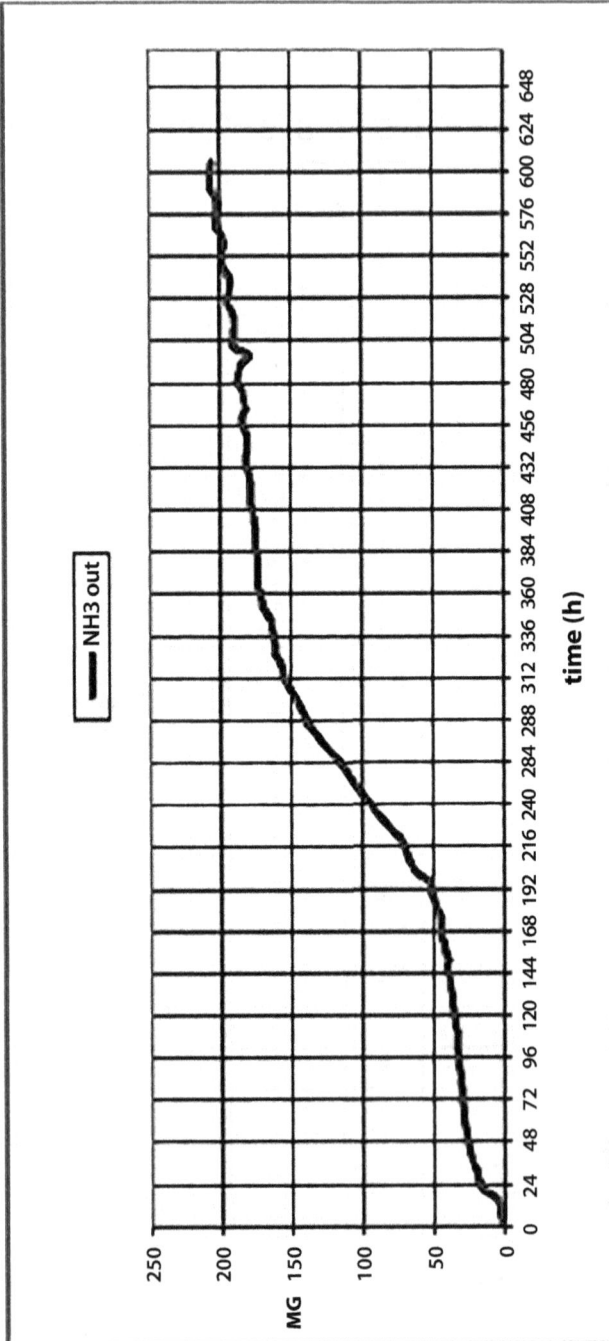

FIGURE 7: Ammonia dynamics during the bio-drying run

Mass loss during the bio-drying run is presented in Figure 4. The mass loss depended strongly on the quantity of putrescible materials present in the input material.

The VS and net water loss were assessed by the quoted model and are presented in Figure 5 (9). The water loss (ΔU) was assessed taking into account the role of hydrogen during the bio-oxidation that characterizes the bio-drying process. VS consumption was minimized thanks to the avoidance of water addition. By this way the presence of carbon was maximized, giving interesting results in term of specific Lower Heating Value (LHV) increase as it can be seen in Figure 6.

The emission factors of ammonia were assessed by modeling and referred to a theoretical case where all the nitrogen present in the treated material is converted into ammonia and emitted through the process air (9). The results are presented in Figure 7.

For a better understanding of the exhausted grape marc use, COD values were generated. To this concern, a part of a sample was dried at 105°C for 24 hours, shredded and homogenized. The COD of the dried exhausted GM resulted 1,961 g_{O2}/kg_{DM}. Another part of the sample was washed and centrifuged at 4000 rpm for 5 minutes. The COD of the washed exhausted GM resulted 1,624 g_{O2}/kg_{DM}. The COD of the liquid phase was determined too. Its value was 27.8 gO_2/l.

The potential rate of microbial self heating of initial and bio-dried exhausted GM were also determined. The obtained results (IR_{24}) were 750 and 500 mgO_2 $kg^{-1}TS$ h^{-1} respectively.

11.4 DISCUSSION

Grape marc, even if exhausted showed a residual activity that supported a quick increase of temperature. Indeed, this phenomenum is confirmed by the fact that at the beginning the temperature in the core of the exhausted grape marc was about 24°C, after only one hour increased and after half day arrived to about 52°C.

Even if the lasting of the process was doubled the results after the first two weeks were suitable instead the results for the last 2 weeks were less important, the process slowed down significantly:

- the temperatures in the last 2 weeks didn't overcome 40°C; on the contrary in the first 2 weeks they reached 55°C (Figure 3);
- the mass, VS and water losses in the last 2 weeks didn't exceed 9%, 9 $g_{(C+H+O+N)}$/kg_{GM} and 61 g_{H2O}/kg_{GM} respectively; on the contrary in the first 2 weeks they arrived to about 23%, 23 $g_{(C+H+O+N)}$ /kg_{GM} and 221 g_{H2O}/kg_{GM} respectively (Figure 4 and 5);
- the LHV increase in the last 2 weeks didn't exceed 15%; on the contrary in the first 2 weeks it reached 27% (Figure 6);
- NH_3 dynamics (Figure 7) shows a potential emission factor into the atmosphere of about 200 mg/kg_{GM} during the whole period; taking into account that the specific flow rate resulted about 10 Nm^3/kg_{GM}, an potential average concentration could be around 20 mg/ Nm^3. This value make it necessary a treatment of the process air.
- Cu content (Table 2) of the bio-dried exhausted GM resulted moderate, in agreement with the content of the original material pointed out in the literature (14). This content is lower than the limit set in the EU for the agriculture use of sludge (15).
- the high concentration of COD in the liquid phase (27.8 g_{COD}/l) can demonstrate that the process of bio-drying is water limited: an addition of water could reactivate the biological activity (16). Anyway the final respirometric index value is typical of a good stabilization.

Concerning the overall process and the obtained results , the optimal lasting for the exhausted grape marc bio-drying could be around 2 weeks.

The measured data from Table 1 and 2 and the modeled data from Figure 5 demonstrate that the used model is reliable in terms of water balance: the final moisture was 52.9% measured versus 53.0% modeled.

The possibility to use the bio-dried material as a SRF for energy proposed was also taken into account in this paper. The top class is expected for Cl (< 50 mg/kg) and Hg (0.022 mg/kg) whilst the final moisture content after bio-drying, being relatively low cannot allow a final LHV typical of the best SRF. This LHV is about 7 MJ/$kg_{biodriedGM}$ after 2 week and respectively 8 MJ/$kg_{biodriedGM}$ after one month. The values are similar to the ones of green wood.

Taking into account all these data, the final product is like an SRF:5,1,1 that is similar with the one obtained from the initial exhausted grape marc: SRF: 5,1,1 (17); the potential microbial self heating for the final product is very low, and low for the initial exhausted GM. Both SRFs-like can be used in a cement factory, but the SRF from bio-drying can be easily stored as its stability is decreased thanks to the moisture limiting content (18).

REFERENCES

1. ISTAT [Internet]. Roma: The Italian National Institute for Statistics, [cited 2012]. Available from http://www.istat.it/it/
2. Devesa-Rey R., Vecino X., Varela-Alende J.L., et al. Valorization of winery waste vs. the costs of not recycling. Waste Manag 2011;31:2327–2335.
3. Fiori L., Florio L. Gasification and combustion of grape marc: comparison among different scenarios. Waste Biomass Valor 2010;2:191-200.
4. Thorngate J.H., Singleton V.L. Localization of procyanidins in grape seeds. Am. J. Enol. Vitic. 1994;45:259–262.
5. Karleskind A. Sources et monographies des principaux corps gras. In: Manuel des Corps Gras. Paris: Lavoisier; 1992. p. 140–144.
6. Maugenet J. Evaluation of the by-products of wine distilleries. II. Possibility of re-covery of proteins in the vinasse of wine distilleries. C. R. Seances Acad. Agric. Fr 1973;59:481–487.
7. UNI/TS 11184: 2006 Waste and refuse derived fuels - Determination of biological stability by dynamic respirometric index, ICS : [13.030.01] [75.160.10], October 2006.
8. CEN/TR 15590:2007 Determination of potential rate of microbial self heating using the real dynamic respiration index, ICS : [75.160.10], October 2007.
9. Rada E.C., Franzinelli A., Taiss M., et al. Lower Heating Value dynamics during municipal solid waste bio-drying. Environ. Technol. 2007;28:463-69.
10. Ragazzi M., Rada E.C., Antolini D. Material and energy recovery in integrated waste management systems: An innovative approach for the characterization of the gas-eous emissions from residual MSW bio-drying. Waste Manag 2011;31:2085-91.
11. Rada E.C., Venturi M., Ragazzi M., et al. Bio-drying role in changeable scenarios of Romanian MSW management. Waste Biomass Valor 2010;1:271-279.
12. Andreottola G., Dallago L., Ragazzi M. Dynamic respirometric tests for assessing the biological activity of waste, Proceedings of the Tenth International Waste Man-agement and Landfill Symposium; 2005 Oct. 3-7; S. Margherita di Pula (CA), Italy.
13. Ciuta M.S., Marculescu C., Dinca C., Badea A. Primary characterization of wine making and oil refining industry wastes. Sci B - Electr Eng 2011;73:307-320.
14. La Pera L., Dugon G., Rando R., et al. Statistical study of the influence of fungicide treatments (mancozeb, zoxamide and copper oxychloride) on heavy metal concen-trations in Sicilian red wine. Food Addit And Contam 2008;25:302-312.
15. 86/278/EEC:1998 Council Directive of on the protection of the environment, and in particular of the soil, when sewage sludge is used in agriculture, 12 June 1986.
16. Rada E.C., Ragazzi M., Panaitescu V., Apostol T. Bio-drying or bio-stabilization process? Sci B Sci B - Electr Eng 2005;67:51-60.
17. UNI 9903-1:2004. Non mineral refuse derived fuels - Specifications and classifica-tion, ICS : [75.160.10], March 2004.
18. UNI CEN/TR 15508:2008. Key properties on solid recovered fuels to be used for establishing a classification system, ICS : [75.160.10], October 2006.

Author Notes

CHAPTER 2

Data Availability

The authors confirm that all data underlying the findings are fully available without restriction. All sequences have been deposited into the NCBI short read archive (SRA) under the accession number SRX484115 for bacteria and SRX485028 for archaea.

Funding

This research has been supported financially by National Key Technologies R&D Program of China (2010BAC67B04) and the key projects of National Water Pollution Control and Management of China (2011ZX07316-004). The funders had no role in study design, data collection and analysis, decision to publish, or preparation of the manuscript.

Competing Interests

The authors have declared that no competing interests exist.

Author Contributions

Conceived and designed the experiments: XD. Performed the experiments: JY. Analyzed the data: BD. Contributed reagents/materials/analysis tools: JJ. Contributed to the writing of the manuscript: JY.

CHAPTER 3

Acknowledgments

The authors wholeheartedly appreciate the financial support offered by Claude Leon Foundation, South Africa.

Conflicts of Interest

The authors declare no conflict of interest.

CHAPTER 4

Competing Interests
The authors declare that they have no competing interests.

Author Contributions
GL and IA designed the experiment. GL and XYZ carried out the experiment. DDF, PGK, and TL participated in the molecular analysis. GL performed the bioinformatics analysis and drafted the manuscript. All the authors have read and approved the final manuscript.

Acknowledgments
This study was funded by the Yangfan project from the Science and Technology Commission of Shanghai Municipality (14YF1400400), the National Natural Science Foundation of China (51408133), and The Danish Council for Independent Research (12-126632) and Strategic Research (12-132654).

CHAPTER 5

Acknowledgments
The authors are grateful to CeLIM NGO for logistical and financial support and to all the people and local bodies for supporting the local activities, in particular CeLIM local staff and Maxixe municipality. This work was part of the "Urban and peri-urban environmental protection: a project for Maxixe Municipality", founded by Comune di Milano and Peppino Vismara foundation.

CHAPTER 6

Funding
DAN received USDA Hatch VT-HO1609MS funding and NF received funding from the National Science Foundation to support this research. The funders had no role in study design, data collection and analysis, decision to publish, or preparation of the manuscript.

Competing Interests

The authors have declared that no competing interests exist.

Acknowledgments

The authors thank Tom Gilbert (HCC), James McSweeney (HCC), and Alexander Utevsky (HCC) for formulating recipes and managing variations of the thermophilic process at a realistic commercial scale. We thank Jessica Henley and Chris Lauber for technical assistance with the laboratory analyses.

Author Contributions

Conceived and designed the experiments: DAN TRW NF. Performed the experiments: DAN TRW. Analyzed the data: DAN JWL. Contributed reagents/materials/analysis tools: DAN NF STB. Wrote the paper: DAN TRW STB JWF NF.

CHAPTER 7

Acknowledgments

The authors wish to acknowledge the R&D collaborative work of UPM-O3 Solutions (2008 to 2010) and Dr. Chon Seng Tan from the Biotechnology Research Centre, Malaysian Agricultural Research and Development Institute, for their assistance and valuable advice.

CHAPTER 8

Competing Interests

The authors declare that they have no competing interests.

Author Contributions

JP: Collected and reviewed the literature and drafted the manuscript. NS: Formulated the objectives, provided guidance and improved the quality of the manuscript. Both authors read and approved the final manuscript.

Acknowledgments

The financial support from Department of Biotechnology (DBT), New Delhi, India, and Department of Science and Technology (DST), New Delhi, India, through Fund for Improvement of Science and Technology Infrastructure in Higher Educational Institutions (FIST), is gratefully acknowledged.

CHAPTER 11

Author Contributions:

Both the authors contributed to conception and design, to acquisition of data and to analysis and interpretation of data; both the authors drafted the article and revisited it critically for important intellectual content; and both the authors gave final approval of the version to be published.

Index

For Product Safety Concerns and Information please contact our EU
representative GPSR@taylorandfrancis.com
Taylor & Francis Verlag GmbH, Kaufingerstraße 24, 80331 München, Germany

www.ingramcontent.com/pod-product-compliance
Lightning Source LLC
Chambersburg PA
CBHW060334220326
41598CB00023B/2705